LUPIN
DEVELOPMENT GUIDE

LUPIN
DEVELOPMENT GUIDE

Miles Dracup

E J M Kirby

University of Western Australia Press

First published in 1996 by
University of Western Australia Press
Nedlands, Western Australia 6907

This book is copyright. Apart from any fair dealing for the purpose of private study, research, criticism or review, as permitted under the Copyright Act 1968, no part may be reproduced by any process without written permission. Inquiries should be made to the publisher.

Any recommendations contained in this publication do not necessarily represent the policy of the Grains Research Development Corporation, Agriculture Western Australia or Centre for Legumes in Mediterranean Agriculture. No person should act on the basis of the contents of this publication, whether as to matters of fact or opinion or other content, without first obtaining specific, independent professional advice which confirms the information contained in this publication.

Copyright © Miles Dracup and E. J. M. Kirby 1996

National Library of Australia
Cataloguing-in-Publication entry:

Kirby, Michael, 1928– .
  Lupin Development Guide.

  ISBN 1 875560 66 1.

  1. Lupins. 2. Lupins – Growth. I. Dracup, Miles, 1959– .
  II. Title.

633.367

Consultant editor: Alex George, Segger's, Perth
Designed by Robyn Tomlinson, Perth
Cover design by Robyn Mundy, Mundy Design, Perth
Typeset in 11.5pt Weiss by Lasertype, Perth
Printed by Scott Four Colour Print, Perth

Contents

Foreword *p. vii*

Acknowledgments *p. ix*

Introduction *p. xiii*

1 The Lupin Life Cycle *p. 1*

2 The Seed and Seedling Emergence *p. 5*

3 Leaves and Leaf Development *p. 13*

4 Branch Development *p. 21*

5 Inflorescence and Flower Development *p. 31*

6 Anthesis and Flowering *p. 37*

7 Pod and Seed Development *p. 43*

8 The Root System *p. 49*

9 Lupin Development Scale *p. 55*

10 Predicting Lupin Development *p. 65*

11 Techniques for Examining Lupins *p. 73*

List of Abbreviations *p. 83*

Glossary *p. 85*

Bibliography *p. 91*

Index *p. 95*

FOREWORD

Traditionally, Old World lupins have been grown as a snack food on a very small scale in the eastern Mediterranean basin. More recently they have been grown in eastern Europe primarily as a fodder and green manure crop, and over the last 50 years they have become increasingly important as a grain legume. The major areas of production are in Poland, the Ukraine and Western Australia. There is now increasing interest in western Europe, the USA and South America in the use of lupins as a grain crop for food and feed.

This interest has developed because farmers have realised there are many benefits in having a grain legume in the rotation. With increasing scientific effort and agricultural production of lupins, it is essential that we have a unified system for the terminology of lupins and description of their development.

As an example of the potential for confusion, the Narrow-leafed Lupin (*Lupinus angustifolius*) in Western Australia has been colloquially called the 'White Lupin'. In the wild state this lupin has blue flowers, but domesticated plants have white flowers with a blue tinge. Varieties of the true White Lupin (*L. albus*) grown in Western Australia are also white flowered with a blue tinge, and to add further to the confusion, there is the Western Australian Sandplain Blue Lupin (*L. cosentinii*). Under these circumstances, when we speak about lupins colloquially, it has been hard to determine which species is being discussed.

Similar confusion may also occur in discussing the development and flowering of lupins when people talk about primary branches and inflorescences, the first-order branches and inflorescences, second-order branches and inflorescences, etc. As well as this branching pattern, we also have types which have restricted or no branching. On occasion these have been called determinant, which technically is incorrect since all lupins are indeterminate.

It is therefore with great pleasure that I write this foreword to a book which has been produced by two scientists well capable of clearing up confusion. I hope that they are not too late and that the rest of us are not too set in our ways to change the way we describe lupin plant parts and development.

I would like to congratulate the authors on a timely and appropriate effort.

Adjunct Professor John Hamblin
Director

Co-operative Research Centre
for Legumes in Mediterranean Agriculture,
The University of Western Australia

ACKNOWLEDGMENTS

Following its conception, development of this *Guide* progressed in fits and starts as the circumstances of the authors changed; for most of the time they were on opposite sides of the globe with many competing demands on their time. It was not until the Co-operative Research Centre for Legumes in Mediterranean Agriculture (CLIMA) began that the path became clear to finish the writing. We are very grateful to Professor Alan Robson, Dr Rodger Boyd and Adjunct Professor John Hamblin at The University of Western Australia for their support during this process.

The *Guide* was compiled from the work and writings of the authors and many other legume scientists. The extensive bibliography of key sources of information is an indication of those whose work has made a major contribution.

CLIMA provided funding and support supplemented with funding from the Grains Research and Development Corporation (GRDC) (formerly the Wheat Research Committee of Western Australia). The salary for Miles Dracup was provided by Agriculture Western Australia. Without the help of these organisations the book would not have been written.

We are also very grateful to the GRDC, Rhône-Poulenc Rural Australia, the Grain Research Committee of Western Australia and the Special Buying Services Rural Independent Agricultural Merchants Association (SBS Rural IAMA), who sponsored the publication.

They say that two heads are better than one. Well, we had the help of several expert heads who generously commented on an early draft and ensured its accuracy. They were: Craig Atkins, John Blake, David Bond,

Wallace Cowling, John Gladstones, Christian Huyghe, Nancy Longnecker, Peter Nelson, Mike Perry, Mark Reader and Brett Thomson.

Ellen Hickman did the excellent line drawings in Chapters 2, 8 and 9 except Figure 8.2 which was done by Jill Ruse. Peter Maloney or Simon Eyres took photographs for Figures 2.2, 5.3b, 7.1a, 8.1, and 11.4, Jon Clements supplied the photograph for Figure 6.6 and Bob Thomson took the photograph used in Figure 11.1. Glenis Ayling kindly proofread the manuscript.

Publication of this book was assisted by the following organizations:

INTRODUCTION

THIS IS A GUIDE for people interested in lupins and lupin production. They may be farmers who are concerned with understanding label instructions for the use of agrochemicals, interpreting advice based on the latest research or assessing and communicating crop growth problems. They may be researchers seeking standard definitions and terminology for exchange of information.

The book brings together a wealth of background information on lupin growth and development. For people new to the crop, this guide shows them what to expect and summarises practical information from many sources. Students can use it as a reference or a laboratory or field manual. Depending on the user, some chapters will be more relevant than others. For example, the development scale chapter (Chapter 9) will probably be most valuable to growers and agronomists, while other chapters such as predicting lupin development (Chapter 10) will probably be more useful to scientists.

The text is fully illustrated with photographs and line diagrams. Specialised scientific terms that may not be familiar are defined in a glossary (Chapter 13).

The *Guide* explains how a particular event is related to the agronomy of the crop or how its timing may be affected by the weather or other factors. Some background information on physiology and/or agronomy, is included for each phase. In addition, although research papers and other sources of information are not cited in the text, the bibliography (Chapter 12) refers readers to important research on lupin development and growth.

Lupins are legumes, that is, members of the sub-family *Faboideae* (see glossary—*'Faboideae'*), which includes several other grain legumes such as field pea, faba bean, and lentil. The family is characterised by the plant's ability to fix elemental nitrogen from the air, and by the fruit, a pod or legume, which contains several large protein-rich seeds. There are some 200 species of lupin, several of which are important crops. Lupins are native to north and south America, Mediterranean Europe and Africa and the tropical highlands of Africa.

The book grew out of the authors' research in Western Australia on Narrow-leafed Lupin (or Narrowleaf or Narrow-leaved Lupin; *Lupinus angustifolius*) and most of the primary descriptions and illustrations are based on this work. Because the development of other cultivated species of lupin (for example, Albus Lupin [or White Lupin; *L. albus*] and Yellow Lupin [*L. luteus*]) is similar to that of Narrow-leafed Lupin, similar principles apply. The book will be equally useful for people working with these species. For example, the explanation of the shoot apex (Chapter 3) would enable someone studying one of the other species to understand its structure and development. Where available, specific information about the other species has been included.

—1—
THE LUPIN LIFE CYCLE

The life cycle of lupins can be divided into phases separated by definable stages. Some of the stages are easily observed (e.g. flowering) but identification of others (e.g. floral initiation) requires the aid of a microscope. Timing of stages and progress through the phases can be predicted by models incorporating climatic data. Anticipation and recognition of developmental stages can be important in crop management where operations need to avoid or target specific phases.

DEVELOPMENT OF THE lupin plant is continuous, from germination to seed ripeness, but definable stages can be recognised, dividing the life cycle into phases, and the rate of development can be measured.

An overall impression of a lupin life cycle is presented in Fig. 1.1, illustrated by the appearance of the plant at intervals. The figure shows the main stages of development (▲) which separate the life cycle into a series of phases, each describing the principal events occurring in the main shoot or branches at that time.

In the following chapters, the various phases and stages are defined and described. Some are easily recognised and form the basis of a growth scale (Chapter 9). This scale precisely identifies significant stages such as seedling emergence

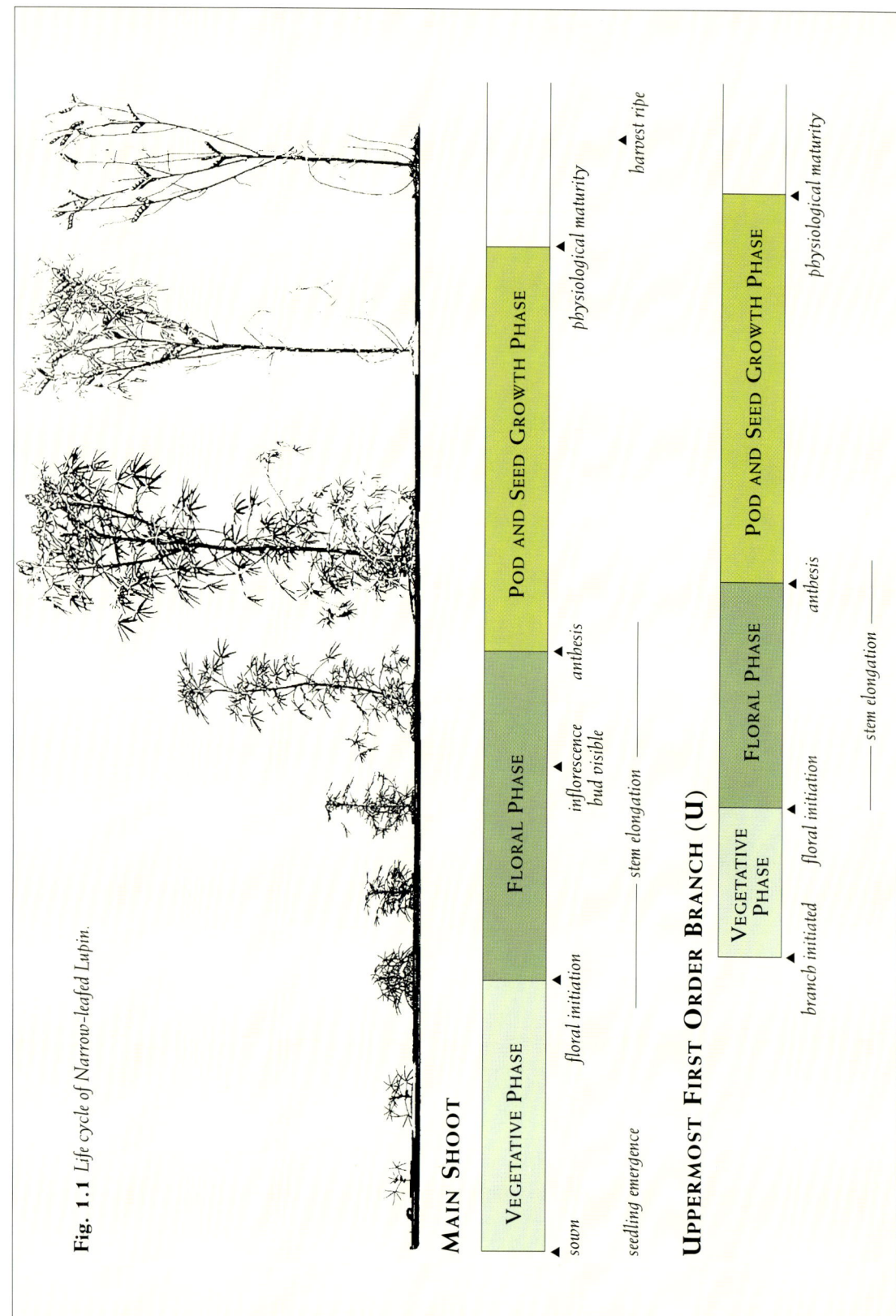

Fig. 1.1 *Life cycle of Narrow-leafed Lupin.*

(Chapter 2) and anthesis (Chapter 6), so that recording or comparing the development of crops can be standardised. Where efficient management depends on a treatment such as an agrochemical application at a particular stage, the scale provides a method of unambiguous communication.

Some critical points in the plant's growth cannot be seen (when they happen) because they occur within the growing point [shoot apex] (Chapters 3 and 5). These changes, for example branch initiation, occur at comparatively early stages of plant development. The final expression of these organs (e.g. branches) depends on the growth conditions during and after their intitation.

Knowing when changes occur at the shoot apex can be used to avoid applying potentially harmful agrochemicals at sensitive stages of development. Equally, this knowledge is important in interpreting the effects of stresses or chemical treatments relative to the phase of development when they were imposed. One such important stage is the beginning of flower formation (*Fig. 1.1; floral initiation*) which can be recognised only by dissecting the shoot apex from the enfolding leaves with the aid of instruments and a dissecting microscope (Chapter 11).

In some phases, particularly pod and seed growth (Chapter 7), colour changes of the pod wall and seed are closely related to ripeness and harvest and form an important element of the growth stage scale.

The rate at which the plant progresses through each phase depends on temperature and daylength. Using defined stages, the duration or rate of progress through a phase can be measured and expressed as a mathematical function. For example, the time from sowing to seedling emergence depends mainly on temperature and, when soil water is adequate, emergence occurs in a certain number of thermal time units. Time from sowing to seedling emergence can be estimated from the ratio [required thermal time/average daily temperature] (Chapters 2 and 10).

The life cycle diagram (*Fig. 1.1*) shows no time scale. The date when a particular stage occurs depends on species, variety, time of sowing and weather, and will vary between different regions. For example, a Narrow-leafed Lupin crop sown in Western Australia late in May flowered (anthesis) in the middle of August, 79 days later. In England, the same variety sown late in March took about the same time to reach flowering; although the temperature was lower, which slowed development, the days were longer which hastened development (Chapter 3).

Information about how plant development and growth are related to the weather is used to build 'models' of lupin growth. Chapter 10 describes a model relating the time of seedling emergence to temperature. The way in which temperature and daylength can be used to predict the time to flowering is also shown. Computer models incorporating growth and development functions can simulate growth and yield. Crop models can help a farmer wishing to determine, for example, flowering date or to assess the tactical economics of a management operation. They can also help a researcher exploring complex interactions between sowing date, variety and weather and thus permit extrapolation beyond the findings of individual experiments.

2

THE SEED AND SEEDLING EMERGENCE

The embryo of a mature seed is a well developed plant, although the root and leaves are still small. An important part of the embryo is the cotyledons which store food to nourish the emerging seedling. The nutrients in the cotyledons also make lupin seeds valuable animal feed. In its dry state, the embryo is fragile and vulnerable to mechanical damage, such as during handling at harvest or sowing. When placed in suitable conditions the seed takes up water and grows. As the shoot passes through the soil, its delicate apex is protected by a hypocotyl hook. The shape of this hook may prevent the seedling emerging in poor soil conditions such as surface crusting. The time taken for emergence depends mainly on temperature and can be measured in thermal time.

AS WELL AS being the basis for the next generation, the lupin seed is the main economic product and is a valuable food source for animals and humans.

SEED STRUCTURE

The seeds of Narrow-leafed and Yellow Lupin are spherical. The seed of Albus Lupin is larger, round but flattened and often dimpled. Both the colour of the seed coat (testa or hull) and the weight of the seed vary among varieties (*Table 1.1*). Narrow-leafed Lupin seeds tend to be heavier than those of Yellow Lupin. Seed stored under normal conditions has a water content of about 10 per cent.

Table 2.1 *Typical ranges of seed mass and colour in domesticated lupins.*

LUPINS	SEED MASS (mg)	SEED COLOUR
L. angustifolius	30–240	*White or cream, sometimes speckled brown*
L. albus	180–950	*White or pink*
L. luteus	70–150	*White, sometimes speckled black*

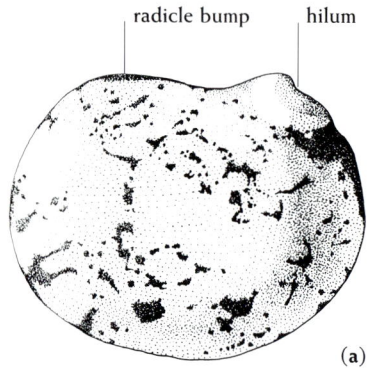

(a)

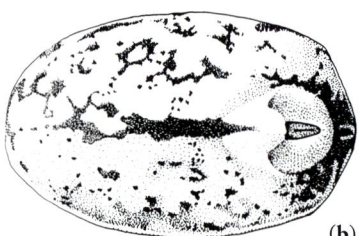

(b)

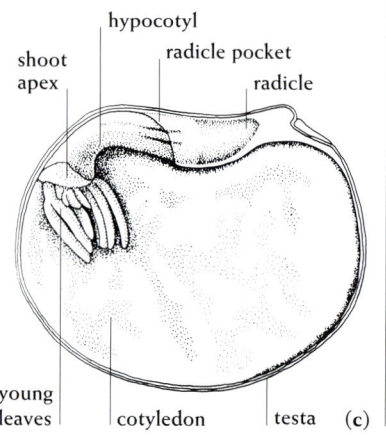

(c)

When Narrow-leafed Lupin seed is viewed lengthways and in profile (*Fig. 1.1*), two bumps are seen. One marks the position of the seedling axis beneath the seed coat, and the other is a distinct scar or hilum (*Fig. 2.1*). The hilum is the scar left when the seed separated from the seed stalk (funicle).

The seed coat (testa) encloses the embryo, which is already a well developed plant. In the dry seed, the coat adheres firmly to the two large cotyledons. The cotyledons are joined to and partially envelop the seedling (or embryonic) axis (*Fig. 2.1*), which consists of the embryo shoot and the radicle or embryo root. The embryo shoot consists of the hypocotyl, the cotyledons and the epicotyl. The cotyledons are attached at the top of the hypocotyl and the epicotyl is the tissue above that (the shoot apex at this stage), with a terminal bud having about five or six leaf primordia (of which four are recognisable leaves) and the meristematic dome or apex. The radicle is inserted in a 'pocket' in the seed coat.

The size and brittle nature of the embryo when dry makes lupin seeds vulnerable to damage during harvesting, drying,

Fig. 2.1 *A seed of Narrow-leafed Lupin seen from the side (a) and top (b). After removing the seed coat, the embryo can be seen. In (c) one cotyledon and part of the seed coat has been removed to reveal the seedling axis. The positions of the hilum and, in (c), the seed coat (testa), radicle, hypocotyl, young leaves, shoot apex and one cotyledon are shown.*

handling and sowing. Either the seedling axis or the cotyledons may be cracked, and the point of attachment between the two is particularly vulnerable. If damaged seeds are sown, establishment and seedling growth may be poor. If a cotyledon is broken off in the seed or badly damaged, less nutrient is available for the seedling. If the radicle is broken there will be little or no root growth. Cracks in the embryo also increase susceptibility to infection by pathogens.

SEED COMPOSITION

The cotyledons account for the bulk of the seed. In Narrow-leafed Lupin the cotyledons are about 72.5% of dry mass, compared with the seedling axis (3.5%); the seed coat is relatively thick (24%). There is no endosperm, the cotyledons being the food store for seedling growth. The seed coat is a smaller proportion of the seed in Albus Lupin (Table 2.2). The lupin seed has a greater proportion of its weight in the seed coat compared with other grain legumes such as faba bean (about 14%), soya bean (about 10%), field pea (about 9%) and kabuli chickpea (about 6%).

Lupins are valued primarily for their high protein content (Table 2.2). Levels of lysine and the sulphur-containing amino acids methionine and cysteine are comparatively low and must be supplemented for some animals feeds. The oil content is generally higher than in other pulse crops and cereals but much lower than in soya bean and oilseeds (Table 2.2). Although containing functionally important oils, oil levels are

Table 2.2. *Typical weights (per cent of total dry mass) of several components of lupin and soya bean seeds.*

	SEED COAT	PROTEIN	OIL	CRUDE FIBRE
L. angustifolius	24	32	6	15
L. albus	18	36	9	10
L. luteus	24	38	5	16
Soya bean	10	34	19	4

considered economically unattractive. Lupin seeds have little starch but are rich in dietary fibre, mainly in the cell wall material of the cotyledons.

Lupin protein and oil are more digestible than those of soya bean, and lupins have low levels of the common anti-nutritional factors. Alkaloid levels vary, but are not a problem in the Australian sweet lupin (Narrow-leafed Lupin).

WATER IMBIBITION

Before a seed can germinate it must first take up water by imbibition. At the time of sowing, about 10 per cent of the mass of the 'dry' seed is water. When the water content reaches about 60 per cent of the imbibed seed weight, germination occurs.

Undomesticated lupins generally have hard seeds (the impermeable seed coat prevents uptake of water when placed in ideal conditions). Hard-seededness breaks down over time with temperature fluctuations so that a proportion of hard seeds will germinate each year. To grow hard-seeded lupins, for example *L. cosentinii* (Western Australian blue lupin), the seed must be scarified. This makes the seed coat permeable and overcomes hard-seededness.

Hard-seededness is genetically controlled. It has been removed by breeding and selection during domestication so that crops with good quality seed have high, uniform rates of germination. Seeds that fall to the ground during harvesting germinate when conditions are suitable and are killed during seedbed preparation so that they do not persist in the soil to pose a 'weed' problem to crops. Domesticated, soft-seeded lupins may still develop hard-seededness if dried to very low moisture content.

SEED GERMINATION

After imbibition, the embryo starts to grow. The radicle grows fastest; its elongation within the seed exerts pressure on the softened seed coat which consequently ruptures near the hilum. When the radicle has emerged and is 5 mm long, the seed has germinated. Under optimum conditions the seed germinates 1–1.5 days after sowing. Well harvested, properly stored soft seeded lupins have a potential germination exceeding 90 per cent.

Seedling Emergence

After germination, the radicle continues to grow downwards and firmly anchors the plant in the soil. This counters the forces exerted as the hypocotyl elongates. At this time the hypocotyl forms a hook and the cotyledons are folded tightly together and point almost downwards (*Fig.* 2.2). Growth of the hypocotyl draws the cotyledons to the soil surface. The hook protects the tender young leaves and the shoot apex from damage by soil abrasion. Eventually the hook reaches the soil surface, sometimes still covered by the ruptured seed coat. This is epigeal seedling emergence. It contrasts with hypogeal emergence, found in faba beans, field peas and some other grain legumes, where the epicotyl rather than the hypocotyl grows and the cotyledons consequently remain below the soil surface.

Fig. 2.2 *Photograph of a young Narrow-leafed Lupin seedling extracted from the soil before emergence to show the hypocotyl hook, hypocotyl and radicle.*

The large projected area of the hypocotyl hook and cotyledons of lupin encounters considerable resistance from the soil to their upward movement. The resistance is less during hypogeal emergence since the cotyledons remain where the seed was placed, and in cereals where only the thin coleoptile emerges through the soil. Lupins are therefore comparatively sensitive to soil crusting which further increases resistance to emergence. There is also greater resistance to emergence of large cotyledons than of small ones.

Seed damage may significantly affect emergence. If the radicle is broken there is no anchorage and the cotyledons cannot be pushed through the soil by hypocotyl elongation. Without the downward growing root, the growth of the hypocotyl lacks direction; it twists and bends and the remains of the root may be pushed towards the soil surface, giving the appearance of a downward growing seedling (silly seedling syndrome).

Fig. 2.3 *Photograph illustrating the definition of seedling emergence.*

The hypocotyl hook is formed by the differential growth of cells in the hypocotyl, leading to longer cells on the outer edge than on the inner edge. Once the seedling emerges and is exposed to light, particularly light in the far-red and blue wavelengths, hypocotyl elongation stops and the hook straightens, bringing the cotyledons up to face the sun. The hook straightens because of continued growth of cells on the inner curve of the hook and arrested growth of cells on the outer. The final length of the hypocotyl therefore depends on the depth of sowing.

A seedling is defined as having emerged when any part of the seedling (that is the cotyledons or hypocotyl) protrudes

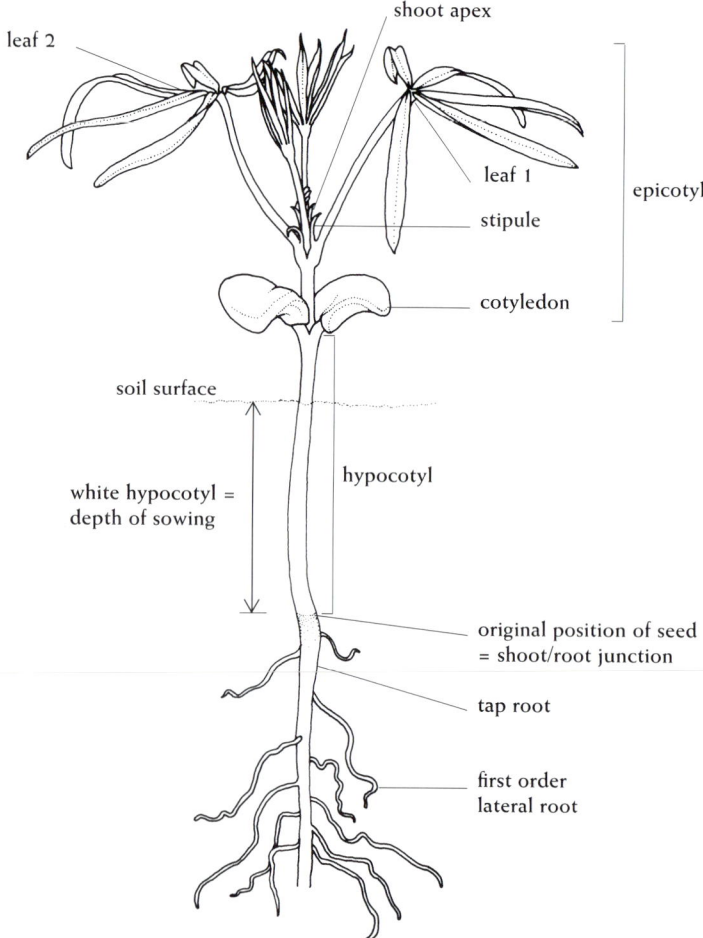

Fig. 2.4 *Diagram of Narrow-leafed Lupin seedling to indicate the positions of various parts.*

through the soil surface (*Fig.* 2.3). [Other definitions of seedling emergence have been used, for example, when the cotyledons are 'clear of soil surface' or 'fully clear of soil and fully open', but these can be ambiguous if, as often happens, the cotyledons do not clear the soil surface.]

A crop is judged to have reached seedling emergence when 50% of the final number of seedlings have emerged.

After emergence, the cotyledons become green and expand and the young leaves grow between them. The hypocotyl above the soil also turns green. The depth of sowing can therefore be determined by measuring the length of white portion of the hypocotyl (*Fig.* 2.4). The junction of white hypocotyl and root can be distinguished by an abrupt change in colour, or, if held up to the light, by a change in the light transmission properties.

Physiology of Germination and Seedling Emergence

Successful germination and seedling emergence depend mainly on an adequate supply of soil water, and suitable temperature, and soil factors such as soil strength.

If soil water and soil strength are not limiting, the rates of germination and emergence depend on temperature and are fastest at about 20°C (*Fig.* 2.5). At 20°C the radicle and hypocotyl elongate at about 2–2.5 cm/day, and when sown at a depth of 4 cm, seedlings emerge in a little over 5 days. In dry, waterlogged, cold, hot or compacted soil, germination and emergence rates and percentage emergence are reduced. Seeds do not germinate when soil temperatures are below 0° or above 30°C or if soil moisture is below wilting point (−1.5 MPa). When crops are sown 'dry' the effective sowing date therefore becomes the date of the first adequate rains.

Emergence is particularly slow in dry soil because the seedling puts more energy into root than shoot growth; this improves the seedling's capacity to search for moisture and also delays emergence, when the seedling starts to transpire water rapidly.

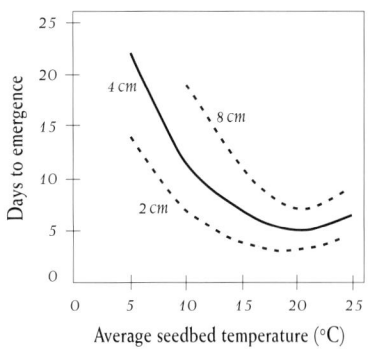

Fig. 2.5 *Emergence of Narrow-leafed Lupin from three sowing depths. Emergence is fastest at 20°C and is delayed at higher or lower temperatures or by sowing deeply.*

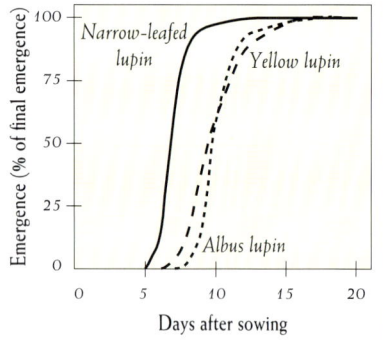

Fig. 2.6 *Graph to show progressive emergence of Narrow-leafed, Albus and Yellow Lupins at near optimum conditions.*

Deep sowing (8 cm or more) will delay emergence and the proportion of seedlings emerging will be reduced; seedlings will be less vigorous and less able to withstand stress. Large seeds appear less affected by deep sowing because of their greater food reserves, which can maintain pre-emergence growth for a longer period.

Narrow-leafed Lupin emerges more rapidly than does Yellow or Albus Lupin (*Fig. 2.6*), but when sown deeply emergence is more rapid in Albus than in the other two species which have far smaller seeds.

Under the fluctuating temperatures of field conditions, time to emergence can be predicted using 'thermal time' techniques. For example, in Narrow-leafed Lupin, when seeds are sown 4 cm deep in moist soil, seedlings can be expected to emerge about 100°Cd (base temperature 0°C) after sowing, that is, in about 6–7 days when the mean daily soil temperature is 15°C.

3

LEAVES AND LEAF DEVELOPMENT

Photosynthesis is carried out mainly by the leaves. Growth of the plant and crop depends on the rate at which the leaf canopy develops. Leaves develop from primordia on the shoot apex until the apex switches to forming flower primordia. Leaves develop within the terminal bud, emerge from it, and continue to grow to take their place in the canopy. The morphology and size of leaves change progressively up the stem. Leaves are spirally arranged on the stem, thus minimising shading of other leaves, and they also move to track the sun. The orientation of leaflets during solar tracking can change to maximise or minimise their irradiation, depending on the plant's water status.

THE INITIALS OF about six leaves are already present in the seed at sowing (Chapter 2). After imbibition, these start to grow and more leaf primordia are initiated. Some time after emergence from the soil, leaves begin to emerge from the terminal bud. Leaf initiation continues until the apex becomes floral and starts to initiate flowers. At this stage the plant has about eight emerged leaves, is still small, dome-shaped and bushy, and about 5 cm high (*Fig. 1.1*). Leaves emerge and grow until the uppermost leaf unfolds, completely exposing the inflorescence (*Fig. 1.1, 'inflorescence bud visible'*).

Unlike cereals, in which the shoot apex remains underground during vegetative growth, after seedling emergence the lupin shoot apex is above the ground (*Fig. 3.1*) and so is vulnerable to physical damage such as sand blasting and grazing or insect

Fig. 3.1 *Photograph of a young Narrow-leafed Lupin plant to show the position of the main shoot apex. The cotyledons and first leaf are also labelled.*

Fig. 3.2 *A diagram of a shoot apex from a young Narrow-leafed Lupin plant to show the apex viewed from above. The scale bar is 0.5 mm. The numbers refer to the leaf position, numbered from the base (leaves 1–14 were removed to expose the apex).*

predation. In seedlings, all buds are easily damaged and the plants cannot recover by producing new shoots.

Development of the Shoot Apex

The shoot apex in a young vegetative plant (2–4 emerged leaves) is a shallow dome of tissue about 0.2 mm in diameter and 0.1 mm high (*Figs* 3.2, 3.3a, b). A binocular microscope is needed to view it, and the surrounding leaves must be removed, as described in Chapter 11.

The apex grows by cell division, and as it increases in size bumps appear on its flanks (*Fig.* 3.2). Each bump is a leaf primordium that eventually becomes a leaf. The first and second pairs of leaf primordia (i.e. leaves 1 and 2, and 3 and 4) are opposite one another, but as the apex grows the primordia become spirally arranged and each successive primordium arises about 137–146° from the previous one (*Fig.* 3.2). The spiral may be clockwise as in Fig. 3.2 or anti-clockwise. Both spirals are equally frequent.

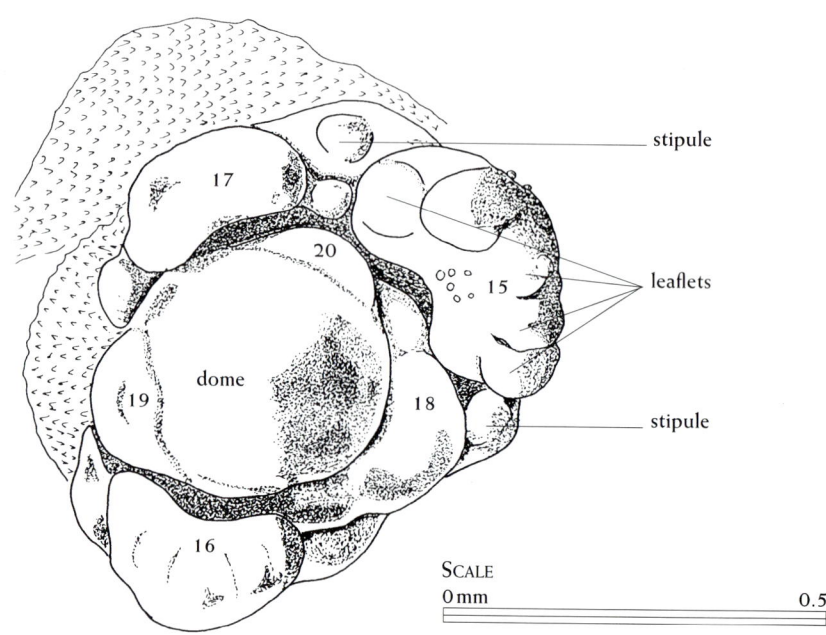

CHAPTER 3: LEAVES AND LEAF DEVELOPMENT

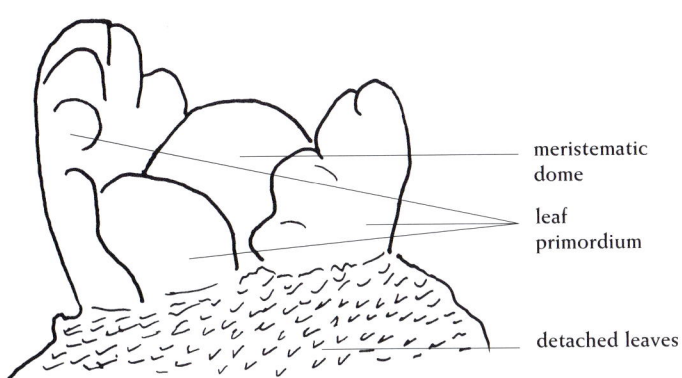

- meristematic dome
- leaf primordium
- detached leaves

Explanatory diagram of (a)

Fig. 3.3 *A dissected terminal bud of Narrow-leafed Lupin: The sequence (a) – (c) shows development in the stages of a leaf.*

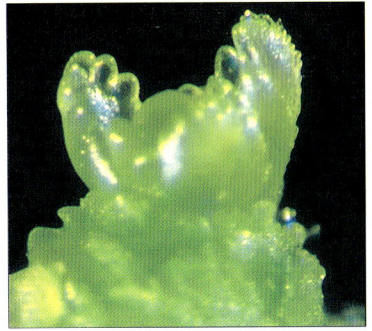

(a) Photograph of a shoot apex and leaf primordia

LEAF DEVELOPMENT

A newly initiated leaf primordium appears as a crescent-shaped bulge on the flank of the apical dome; stipule primordia form at each end of the crescent and the middle part forms a primordium which will become the petiole (leaf stalk) and the leaflets (*Fig.* 3.2). In the first-formed leaves, this middle part then differentiates into five (sometimes four) protuberances which will develop into the leaflets. In the upper leaves, up to nine protuberances differentiate and become leaflets. In the early stages, the leaflets look like fingers of a hand closing over the shoot apex (*Fig.* 3.3a, *explanatory diagram*). At an early stage the cells which later grow into hairs can be seen on the 'palm of the hand' and later all over the primordium (*Fig.* 3.3b, c). The young leaves and the enclosed shoot apex (leaf primordia and meristematic dome) constitute the terminal bud.

(b) Developing leaf primordia

The leaflets and the stipules (small structures at the base of the petiole) continue to grow. When the leaf is about 0.5 mm long it is densely clothed in hairs so that, even using a microscope, it is difficult to distinguish the leaflets. During this early phase the leaflets grow more rapidly than the petiole, which remains short. When the leaf is about 5 mm long, it can be seen clearly on the terminal bud (*Fig.* 3.3c). It remains in the bud until it is about 20 mm long, at which time it has become the outermost

(c) Young leaves on the terminal bud

Lupin Development Guide

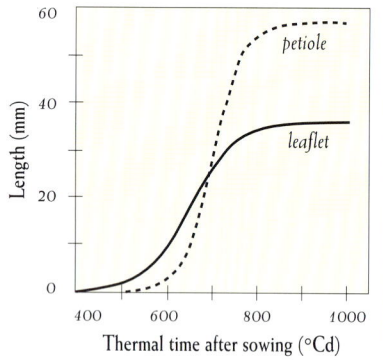

Fig. 3.4 *Graph of the lengthening of the central leaflet (———) and the petiole (- - - -) of a leaf from about three-quarters of the way up the main stem.*

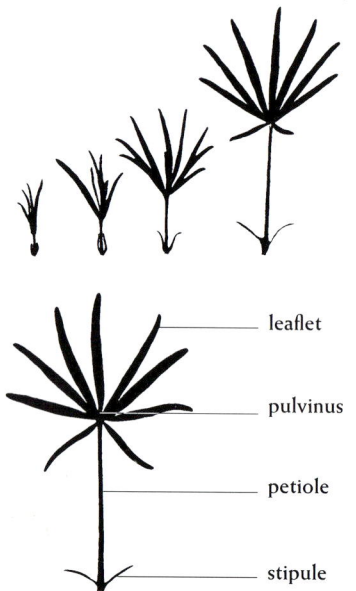

Fig. 3.5 *Stages in leaf emergence and expansion in Narrow-leafed Lupin. The upper left-hand silhouette illustrates a leaf that has just emerged.*

leaf in the bud. The leaflets then begin to spread out and unfold about the main vein and the petiole grows rapidly (*Figs* 3.4, 3.5). This is when the leaf is said to be emerging from the terminal bud. As the leaf grows, the petiole inclines outwards and the leaflets continue to expand and unfold until, at maturity, the leaf becomes more or less horizontal and takes its place in the canopy.

The shoot apex remains more or less unchanged in size and shape until the final leaf is initiated. The shoot apex then switches from producing leaf primordia to flower primordia (floral initiation), the apex becomes larger, and the height of the dome increases relative to the diameter.

RATES OF LEAF INITIATION, EMERGENCE AND FINAL NUMBER OF LEAVES

The rate at which leaves are initiated may be determined by making frequent counts of the number of primordia on the apex. The most recently initiated primordium is counted when it is clearly visible as a bulge on the smooth flank of the dome. Eventually the topmost primordium will be that of a flower rather than a leaf (*Fig.* 5.1).

Similarly, the rate of leaf emergence can be estimated by frequent counts of the number of emerged leaves. An emerged leaf is defined as one in which the leaf has separated from the terminal bud and the leaflets have begun to diverge from one another (*Fig.* 3.5). The number of emerged leaves increases until the last leaf diverges from the terminal bud.

A leaf lives for only a certain time after which it senesces, that is, it becomes yellow, forms an abscission layer and falls, leaving a scar. The oldest leaves fall first, while the plant is still growing. The rate of leaf senescence increases as the seeds approach their maximum growth rate. By the time of physiological maturity almost all leaves have fallen (*Fig.* 1.1: *physiological maturity stage*).

The final number of leaves on the main stem may be counted before or after the leaves have fallen; after leaf fall, leaf scars can be counted, bearing in mind that the two cotyledons also leave scars.

Rates of leaf initiation and emergence are related closely to temperature and can be estimated by regression of the number of leaf primordia or emerged leaves, respectively, on time or thermal time. For field-grown plants that experience seasonally varying temperatures, thermal time can be used to explain the effect of fluctuating temperatures and is therefore the more informative.

In Narrow-leafed Lupin grown in Western Australia, the time between initiation of successive leaves (plastochron) on the main stem is about 26°Cd, or about 2.5 days at an average temperature of 10°C. Long days increase the rate of leaf initiation and, under the longer spring days in England, this gap is only 18°Cd.

Leaves are initiated more slowly in Albus Lupin. In France, leaves of winter types sown in autumn are initiated about 37°Cd apart, and for spring types sown in spring the leaves are initiated about 25°Cd apart (the base temperature for calculating thermal time is 3°C for Albus compared with 0°C for Narrow-leafed Lupin). Longer days during leaf initiation in spring-sown plants compared with autumn-sown plants could explain the apparently faster leaf initiation in spring Albus Lupin.

In Western Australia, successive leaves of Narrow-leafed Lupin typically emerge (phyllochron) 33–42°Cd apart, corresponding to about 3–4 days at an average temperature of 10°C. As with leaf initiation, leaf emergence is hastened by long days, and in England, under long spring days, the gap between emergence of successive leaves is reduced to 25°Cd. Nutrient deficiency or water deficit can delay leaf emergence.

Rates of leaf emergence for Yellow Lupin are similar to those for Narrow-leafed Lupin and are about 50 per cent faster than for Albus Lupin grown in the same environment. When sown in spring in France or England, leaves of Albus Lupin emerge about 36–45°Cd apart.

The final number of leaves on the main stem ranges from 8 to 40 or more, depending on cultivar and environment,

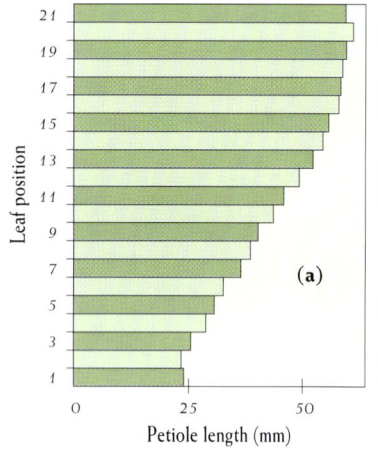
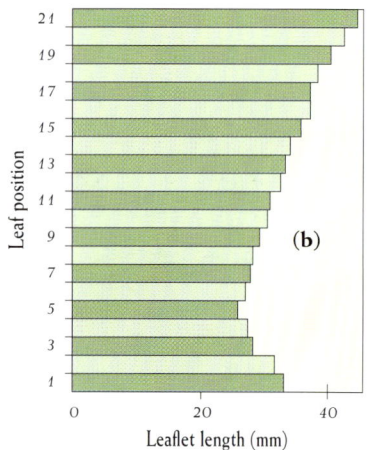
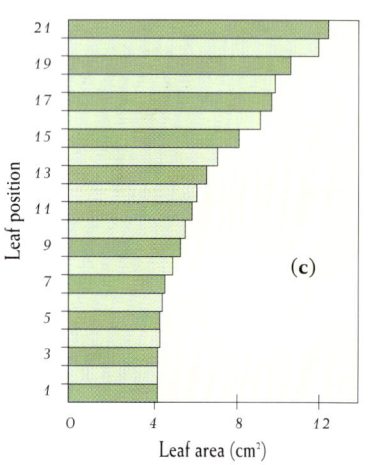

particularly temperature and daylength. Some cultivars require vernalisation, that is, they must be exposed to low temperature (between 1°C and 14–17°C) for a period of time before the shoot apex switches from initiating leaves to flowers.

The period of low temperature needed to reach full vernalisation depends on the cultivar. It also varies with daily temperatures (hence sowing time) and determines the number of leaves on the main stem. The vernalisation requirement can be quantified as discussed in Chapter 10. Lupins are long-day plants, and since long days induce earlier changeover from leaf to flower initiation fewer leaves are formed. Cultivars differ in their response to vernalisation and daylength, and these are important factors in the adaptation of a cultivar to a region and time of sowing.

In northern Europe, winter Albus varieties require low temperatures to form an optimum number (25–30) of leaves. This adapts them to sowing over a relatively long period in autumn; if sown in spring (March) they produce an excessive number of leaves and do not flower or fruit satisfactorily because of the long time taken to reach vernalisation.

Generally, modern Western Australian cultivars of Narrow-leafed Lupin do not require vernalisation. For optimum yield they must be sown during a relatively narrow period; their growing season is dictated by the onset of winter rains followed, after flowering, by rising temperatures and dry spring weather. They do not need a vernalisation response to modify their life cycle for variation in sowing time. They are sown in autumn (April–May) and produce abut 20 main stem leaves. If sowing is delayed until late June there will be one or two fewer leaves on the main stem in response to the lengthening days after seedling emergence.

Fig. 3.6 *Mature size of each leaf on the main shoot of Narrow-leafed Lupin. Leaf 1 is the basal leaf; (a) length of petiole; (b) length of middle leaflet; (c) area of leaf.*

Water deficit and nutrient deficiency can also reduce the final number of leaves.

STRUCTURE OF A MATURE LEAF

The leaf is palmate (*Fig. 3.5*). The leaf base clasps the stem and on either side of the petiole is a narrow, pointed stipule. The mature petiole is longer than the leaflets and is bent at the base so that the leaf is inclined almost at a right angle to the stem. In Narrow-leafed Lupin the leaflets are narrow (broader in other species) and taper to a point. Where the leaflet is attached to the petiole there is a small area, the pulvinus, which is lighter green. In the axil of the leaf there is a lateral bud that may grow out to form a branch.

LEAF SHAPE AND SIZE

In Narrow-leafed Lupin the lowest seedling leaves on the main stem have five (sometimes four) leaflets; the number increases up the stem, reaching nine in the uppermost leaves. Petiole and leaflet length (*Figs 3.6a, b*), leaflet width and leaf area (*Fig. 3.6c*) all increase from the lowest to the uppermost leaf.

Fig. 3.7 *A plant with 20 emerged leaves, viewed from above. The diagram on the right shows the spiral pattern of leaf arrangement which minimises shading of one leaf by another. Leaves 13 down to 1 have shorter petioles and are often obscured by those above.*

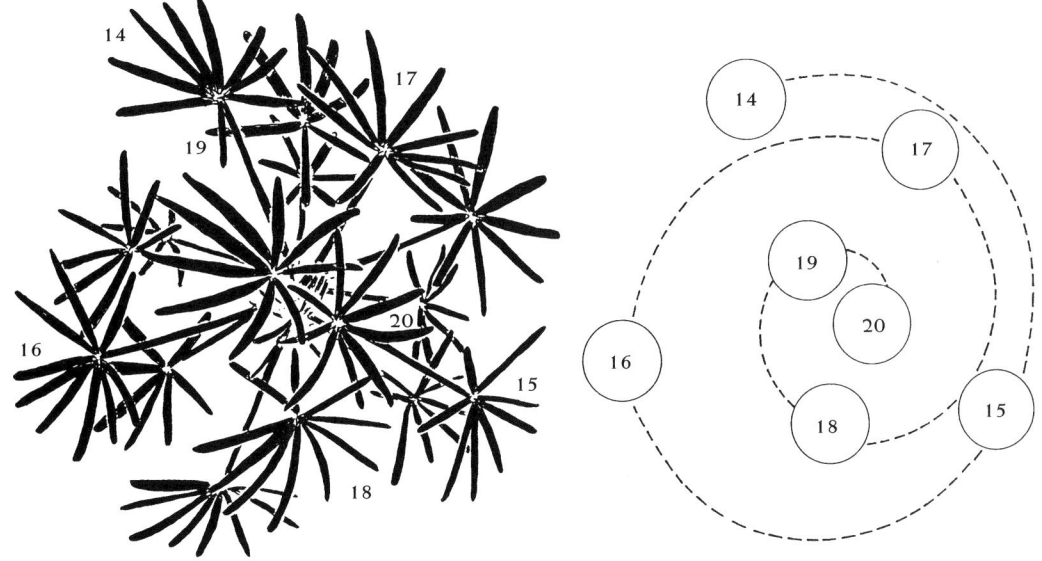

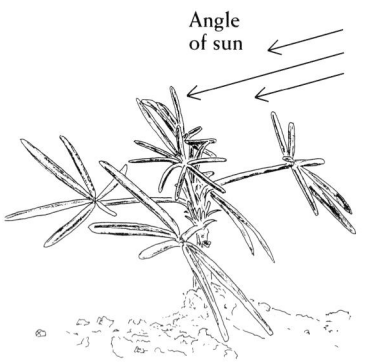

Fig. 3.8 *Diagram of a Narrow-leafed Lupin seedling with leaves oriented to the sun.*

Leaf Arrangement

The first two pairs of leaves are opposite and decussate (arising at the same level on the stem, at 180° to each other). As more leaves form, the arrangement changes until each arises at about 146° to the previous one in Narrow-leafed Lupin and at 137° in Albus; and the internode between elongates resulting in a spiral arrangement (*Fig.* 3.7). Thus, as the plant grows and the crop achieves complete ground coverage, the leaf arrangement minimises shading of one leaf by another and maximises use of light.

Leaf Movements

Throughout the day, fully expanded leaves track the sun (heliotropism). Leaflets may be oriented perpendicular to the sun's rays (diaheliotropism) or be parallel or at an angle to them (paraheliotropism) (*Fig.* 3.8). During the night, the leaves turn so that by next morning they face the rising sun again.

This has important consequences for the level of radiation on the leaf. Solar tracking may maximise or minimise the irradiation of the leaf depending on leaf angle to the sun. Leaf movements are controlled by turgor changes in the pulvinus.

The inclination of the leaf depends on leaf water status. As the leaf water potential declines, so the angle of the leaf to the sun decreases. At high water deficits, leaves are held almost parallel to the sun's rays and the leaflets fold together about the midrib. With ample water, solar tracking enables the plant to maximise the interception of direct solar radiation. If drought conditions prevail, the ability to reduce the level of irradiance lowers leaf temperature, reduces transpiration and increases water use efficiency.

—4—
Branch Development

There are three types of first-order branches on the lupin main stem—upper (or apical), mid, and lower (or basal). The upper branches, which arise in the axils of the three or four leaves beneath the inflorescence, are the most important and usually contribute most to crop yield. They begin development a little after floral initiation on the main stem. Second-order branches develop in axils of the upper leaves of first-order branches and the sequence continues for further orders of branching. Branches of successive orders have progressively fewer leaves, lower vigour and are shorter. Lupins with restricted branching are being studied; they have different branch development and fewer orders of branching.

BRANCHES ARISE FROM lateral shoot apices which form in nearly all leaf axils (the angle between the stem and the base of the leaf petiole). The branch apex is essentially similar to a main shoot apex, and produces leaf and flower primordia, although not as many as the main shoot. The stages of leaf and flower development on branches are identical to those on the main shoot.

The final structure of a branch depends on its position on the stem. Rates of leaf initiation on the branches (*Fig. 4.1*), and total number of leaves on the branches (*Fig. 4.2*) show gradients up the main stem. At some branch positions, fewer leaves emerge from the bud than have been formed (*Fig. 4.2*) and branch apices sometimes do not develop at lower-mid

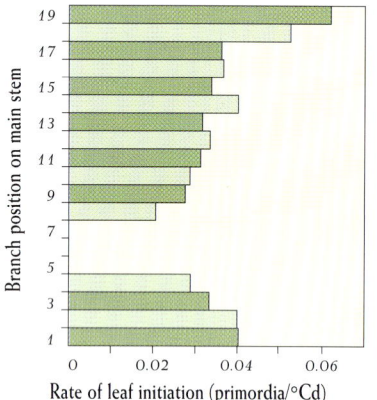

Fig. 4.1 *Branch leaves of Narrow-leafed Lupin (grown in England) initiate faster on branches at basal and apical positions on the main stem than at mid-positions.*

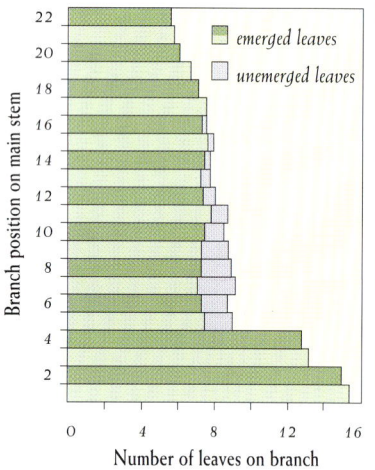

Fig. 4.2 *In Narrow-leafed Lupin most leaves are initiated on basal branches and fewest on apical branches. However, even for the well spaced plants used here, several leaves commonly fail to appear from the bud of branches at mid-positions on the main stem.*

positions on the stem (leaf positions 5–7 in Narrow-leafed Lupin, *Fig. 4.1*).

Because of these gradients, three types of branches, classified by their size and position on the stem, can be recognised. Basal or lower branches arise from buds in the axils of the cotyledons and the lower leaves (generally leaves 1–4); upper or apical branches arise from buds in the axils of the uppermost (generally 3 or 4) leaves. In between the basal and upper branches, in the mid part of the stem, are small branches. The occurrence, amount of development and number of each type of branch depends on factors such as plant population density and mineral nutrition.

UPPER BRANCHES

The branches that arise from the axils of the uppermost leaves on the stem (*Fig. 4.3*) grow strongly and generally produce an inflorescence and productive pods. As the branches grow longer and their leaves expand, they over-top and shade the main shoot leaves and inflorescence.

The number of leaves on the upper branches of the main shoot typically increases from five on the uppermost branch to seven on the third branch from the top (*Fig. 4.2*). The upper branches themselves form branch buds in their leaf axils. Some of these 'second-order' branches (see Nomenclature, below) may themselves flower and produce pods and may in turn bear 'third-order' branches. Vigour declines with successive orders of branching; upper branches on the main shoot are always strongest, with largest leaves and longest internodes. These upper, first-order branches are also generally the main contributors to yield. The uppermost first-order branch produces more pods and seeds than branches below it and is usually also more productive than any second- or third-order branch.

BASAL BRANCHES

Basal branches arise from the axils of the lowest leaves of the main shoot and can be the longest of all the branches. They do not occur in Albus Lupin and the following description applies only to Narrow-leafed Lupin (*Fig. 4.4*). Under normal crop management few basal branches produce viable inflorescences or set pods but the basal branch in the axil of leaf 1 is the most likely to produce an inflorescence, and vigour of basal branches declines with each successively higher branch position. Basal branches may also be produced in the axils of the cotyledons.

The mid-stem branches, that is those between the three or four basal branches and the upper branches, generally produce only a few emerged leaves and rarely an inflorescence; usually the branch shoot apex aborts at an early stage of development.

NOMENCLATURE

Branches coming from the main stem are first-order branches. Those arising on the first-order branches are second-order branches and so on. The progression rarely carries on further than fourth-order branching.

Naming and numbering of branches is complicated. One method is to identify them by reference to the subtending leaf, but generally this is not feasible for the upper branches because the number of main stem leaves varies both within and between crops, environments or management treatments. The system, however, is appropriate for the basal branches, which are numbered according to the subtending organ (cotyledon or leaf). An alternative system is used for the upper branches which are numbered with reference to the uppermost leaf, in sequence down the main shoot or branch.

The prefix abbreviation U is used as shorthand to identify the Upper branches. Thus the branch arising in the leaf axil immediately below the inflorescence is U and the one below that, U-1. A second-order branch in the axil of the uppermost leaf on branch U is U/U while that in the uppermost leaf axil of branch U-1 is U-1/U (*Fig. 4.5*).

For the lower, Basal branches the convention is to use the prefix B, plus the number of the subtending leaf or cotyledon; that is a branch in the axil of leaf 1 on the main shoot is denoted B1. In the event that this branch produced a branch in

Fig. 4.3 *Photograph showing upper branches of Narrow-leafed Lupin. The main shoot inflorescence is almost at anthesis and the inflorescence bud of the uppermost branch can be seen.*

Fig. 4.4 *Narrow-leafed Lupin plant with basal branches. B1 and B2 are the branches in the axils of leaf 1 and 2, respectively.*

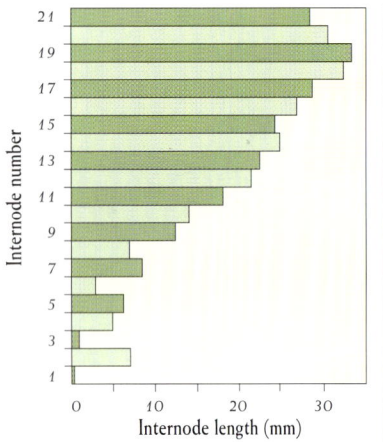

Fig. 4.6 *Lengths of main stem internodes of Narrow-leafed Lupin.*

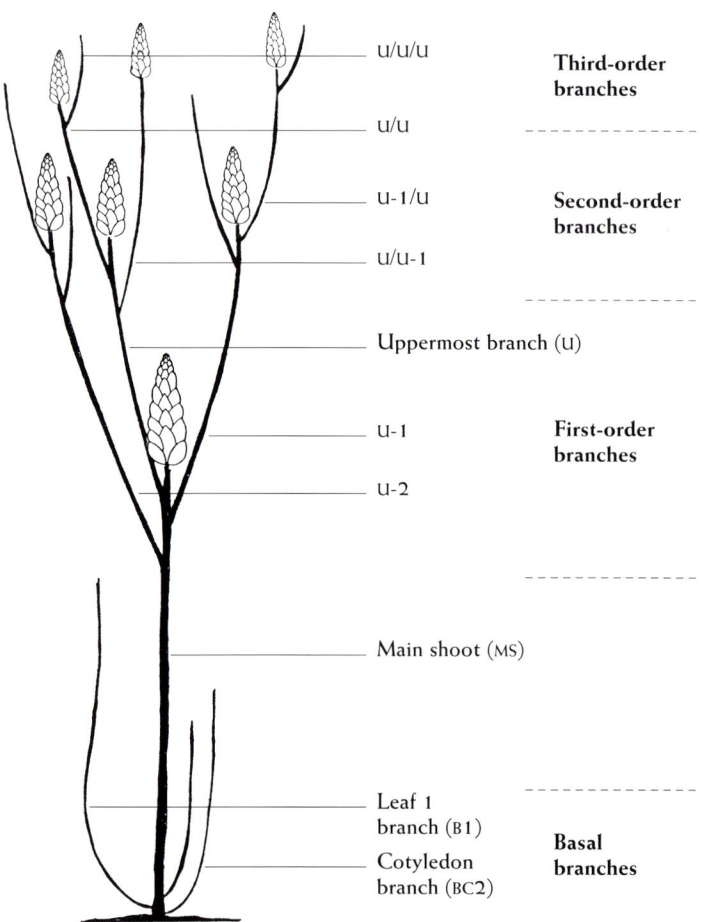

Fig. 4.5 *Diagram to illustrate the system for identifying branches (see text for explanation).*

the axil of its uppermost leaf, it would be designated B1/U. The branches arising in the axils of the cotyledons are denoted BC1 or BC2, although careful inspection of the pattern of leaf arrangement is needed to determine which of the cotyledons is which.

STEM AND INTERNODE GROWTH

At floral initiation the plant is 5–10 cm high and conical in shape, with comparatively little separation between the nodes. After the onset of floral initiation the main stem grows rapidly.

The internodes contribute to stem growth in sequence, starting from the bottom.

Internodes are conventionally named or numbered according to the node from which they arise. Thus the internode above the cotyledons is the cotyledonary internode; the internode between leaves 1 and 2 (of negligible length) is internode 1 and so on.

The first few internodes are alternately long and short. Then, starting at mid-positions, the internodes become successively longer, except the highest few internodes which get shorter (Fig. 4.6). Final stem length is achieved shortly after anthesis. Lupin plants grow taller under short days than under long days, and in Western Australia the main stem of Narrow-leafed Lupins sown near to the end of June is longer than for plants sown earlier or later. Plants are also taller in dense crops than in sparse ones.

After the inflorescence bud becomes visible the peduncle and rachis grow strongly, reaching maximum length some time after the stem (Fig. 4.7).

The upper first-order branches begin to lengthen one to two weeks before anthesis on the main shoot and reach maximum length together. Despite coming from successively lower positions on the main stem the upper first-order branches rise to the same height in the canopy, due to generally longer and more internodes on the lower than on the higher branches.

As with first-order branches, second- and then third-order branches begin lengthening shortly before anthesis on the previous branch order. Branches over-top branches of the previous order but become successively shorter up the plant (Fig. 4.8).

Sequence of Branch Development

The meristem of an upper branch is distinguishable within a few days of the initiation of its associated leaf (Fig. 4.9a). The apices of branches U-3, U-2, U-1 and U arise in sequence.

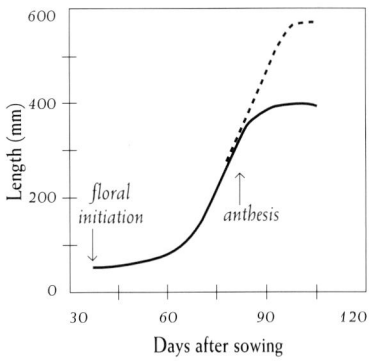

Fig. 4.7 *Length of the main stem (———) and the stem plus peduncle and inflorescence (- - - -) versus days after sowing of Narrow-leafed Lupin.*

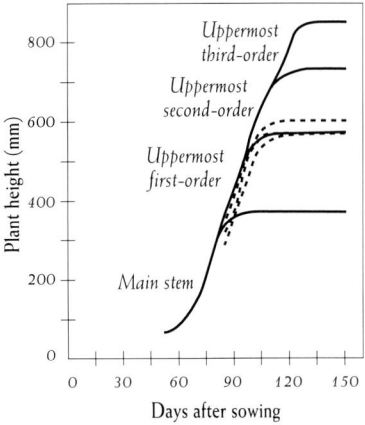

Fig. 4.8 *Contributions of the various stems of Narrow-leafed Lupin to plant height. The dashed lines are for the first-order branches, U-1, U-2 and U-3.*

Lupin Development Guide

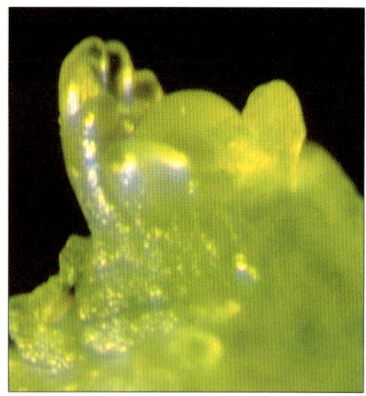

(a) Early development of an upper branch

(b) Later stage of branch development showing its shoot apex

(c) Photograph of complete plant at the same stage of growth as in (b)

Fig. 4.9 (a) Photograph of an early stage of development of an upper branch of Narrow-leafed Lupin; explanatory diagram of (a) to show the meristem of the axillary branch U-2, two leaf primordia and the dome of the apex of the main shoot shortly after floral initiation; (b) photograph of a later stage of development of an upper branch, showing the shoot apex of branch U; the bud of the main shoot inflorescence is covered in hairy bracts; (c) photograph of the plant at the stage of branch development shown in (b).

Explanatory diagram of (a)

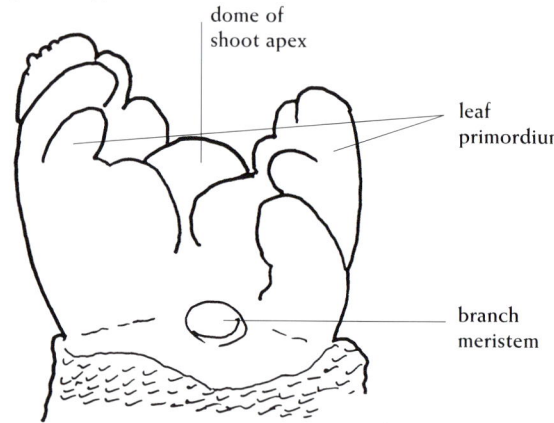

By the time the main shoot inflorescence is about 3 mm long, the bud of branch U has three or four visible leaf primordia (Fig. 4.9b) and the plant is still small (Fig. 4.9c). Differences in the final number of leaves on the branches and small differences in the rate of leaf initiation mean that flowers are initiated first on U and that this is the first upper branch to reach anthesis (Fig. 4.10). At about the time of floral initiation on the branches, the main shoot inflorescence ceases to initiate flowers (compare Figs 5.2 and 4.10) and there are about 14 emerged leaves on the main shoot. Similar relationships between flowering and branch growth exist between first- and second-order branches and between second- and third-order branches.

MODIFIED BRANCHING PATTERNS

Pod set and growth on the main shoot, thickening of the main stem and vigorous growth of the first-order lateral branches all

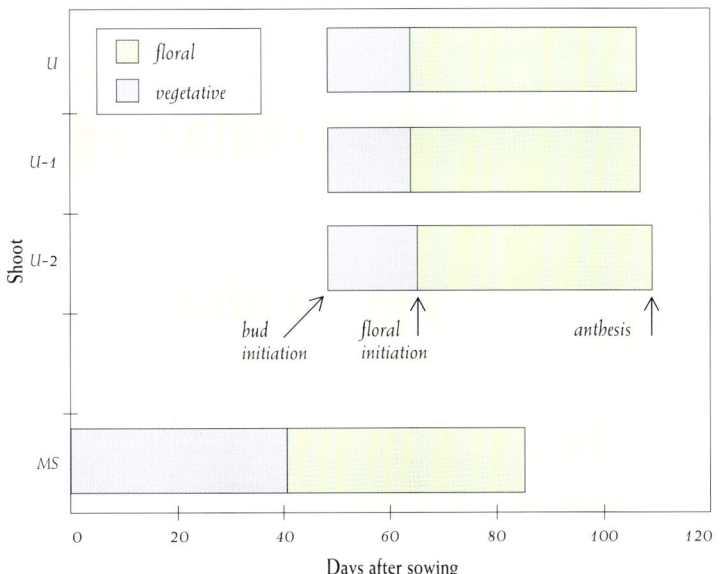

Fig. 4.10 *Life cycle diagram to show the relation between the development of the main shoot (MS) and the upper first-order main shoot branches in Narrow-leafed Lupin. The leaf initiation phase extends from sowing or branch bud initiation until floral initiation and the floral phase extends from floral initiation until anthesis. There were 21 leaves on the main shoot, 5 on U, and 6 on U-1 and U-2.*

occur at the same time. It has been suggested that this generates competition for resources (assimilates, mineral nutrients or hormones) between the main shoot and branches. Also, the lateral branches grow above the main shoot and shade its leaves and inflorescence. Competitive effects of these lateral branches have been demonstrated by cutting them off at an early stage of development; the main shoot inflorescence then sets more pods and produces a greater yield.

Many of the branches formed by the lupin plant produce pods and, in general, more pods are set than will grow and fill seeds to maturity. Some physiologists believe that such over-production of branches is a wasteful process since it promotes excessive competition within the plant for resources, shades organs low in the canopy and leads to high water use (transpiration) at a critical time in the plant's life cycle, hastening senescence.

Because of these hypotheses, physiologists and plant breeders are exploring the possibility of finding treatments or breeding genotypes with less branching. There are promising breeding

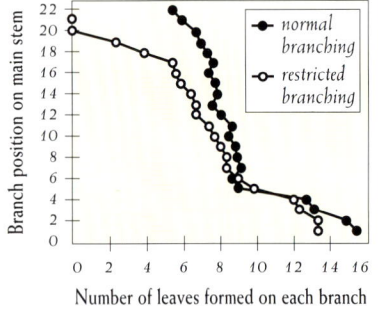

Fig. 4.11 *Number of leaves on main stem branches of restricted branching and conventional branching genotypes of Narrow-leafed Lupin. Branches are numbered from the base upwards.*

programs which aim to develop cultivars with restricted branching and higher yields.

A restricted branching genotype, as the name implies, is recognised by less profuse branching, particularly of the second and higher orders. Because branching is strongly influenced by environment and management, however, this does not provide a satisfactory definition of restricted branching. Restricted branching plant types differ most importantly from conventional types in producing fewer (*Fig. 4.11*) but larger leaves on the upper branches. This reduction is most marked in long-day environments, such as plants grown over summer in Europe.

In some genotypes the number of leaves per branch may be reduced to zero and only a flower or a small inflorescence forms in the uppermost leaf axils of the main shoot (*Fig. 4.12*). With fewer leaves and orders of branching there is less shading of pods (*Fig. 4.13*), fewer branches which do not produce pods and a shorter period of flowering, pod and seed growth on the main shoot and branches.

Yield comparisons of restricted branching and normal types of both Narrow-leafed and Albus Lupins have shown that the restricted branching types have the potential to out-yield conventional types which have otherwise similar genetic backgrounds.

NOMENCLATURE FOR TYPES OF RESTRICTED BRANCHING

The terminology used at present to describe the morphology of the different branching types is unsatisfactory. Descriptions of lupin morphological types use different terms; for example, plants that produce a main shoot only are variously referred to as 'determinate', 'fully reduced branching' or 'epigonal'. Plants that branch freely are called 'indeterminate'.

In authoritative botany textbooks a 'determinate' inflorescence is one in which the main shoot apex becomes a single flower and further flowers are produced on lateral branches beneath it (this type of inflorescence is a cyme, see Glossary).

'Indeterminate' inflorescences have an apex that continues to produce flowers, and flowering usually starts at the bottom of the inflorescence (for example, a raceme, see Glossary). Both restricted branching and conventional lupins produce a raceme, so that by the classical definition they are indeterminate. They differ in the number of leaves formed on each branch, which reduces the potential for the production of higher order branches.

A further problem is that some of the descriptive terms used to describe lupin morphology do not appear in either standard or specialist biological or botanical dictionaries and rigorous definitions are not available (for example, the term 'epigonal').

A more rational method to describe the various morphological branching types is by reference to the upper lateral branching on the main shoot.

- Branching can be restricted because of few leaf nodes on the upper first-order branches, and, in turn, on the second-order branches. This type of branching may be referred to as **mildly restricted**, and in Narrow-leafed Lupin grown in Western Australia it can be conveniently recognised by the uppermost first-order branches having only 1–4 leaves compared with 5 or 6 in normal branching types.

- Branching can be **restricted** because the plant fails to produce branches at one or more upper leaf nodes on the main shoot. This type of branching may be referred to as restricted and in place of a branch in the axil of upper leaf nodes a range of structures may be found, from nothing to a single flower or pod (for example, Fig. 4.12), to a small raceme on a peduncle.

- Branching can also be **fully restricted** by there being no branch in the axil of any leaf on the main shoot. Instead, a range of structures similar to those described above is found in the axils.

Fig. 4.12 *Photograph of a restricted branching Narrow-leafed Lupin plant showing upper leaves with flowers in their axils.*

Fig. 4.13 *Photograph of restricted branching (left) and normal branching (right) Narrow-leafed Lupin plants to show the difference in exposure of the main shoot inflorescence.*

Previously, the terms determinate, semi-determinate, reduced, and epigonal (= fully restricted) have been used to describe restricted branching states.

Plant morphology is not definitive for a cultivar. It varies according to location, plant spacing, nutrition and sowing date. For example, fewer leaf nodes are produced on first-order branches of Narrow-leafed Lupin grown under long days (such as in summer in Europe) than under short days such as in winter in southern Australia. A range of morphologies therefore may be found within and between environments, fields or plots.

—5—
INFLORESCENCE AND FLOWER DEVELOPMENT

The floral phase starts with floral initiation, when the main shoot apex changes from producing leaf primordia to producing flower primordia. Flower buds then accumulate on the shoot apex forming an inflorescence and each develops floral organs. Meanwhile, the main stem leaves and stem grow vigorously and the upper, first-order branches which contribute substantially to the final yield are initiated and begin to grow. Eventually flowers are mature and ready for anthesis. Cultivar adaptation depends largely on the timing of this phase. High yields depend on healthy crops which grow and develop well during this phase.

ALTHOUGH LEAVES CONTINUE to appear on the main stem until the inflorescence becomes visible, leaf primordia are initiated only for a relatively short time (*Fig. 1.1*). Thereafter, a major change occurs as leaf initiation ceases and the meristematic dome reorganises to form flower initials. This important event is called floral initiation. Following floral initiation, many flowers are initiated in sequence over several days and form an inflorescence. Leaves, initiated in the vegetative phase, continue to grow, emerge from the terminal bud and expand. This, together with stem growth, eventually exposes the young inflorescence. Later, petals on flowers unfold, indicating the onset of the next phase, anthesis (Chapter 6). Development and growth of the upper lateral branches, which also produce inflorescences, occurs during this phase of flower development (Chapter 4; *Fig. 1.1*).

Lupin Development Guide

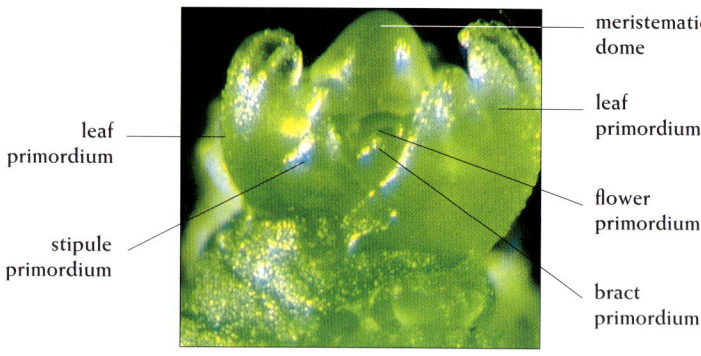

Fig. 5.1 *Shoot apex at floral initiation stage in Narrow-leafed Lupin. The conical dome has a flower primordium subtended by a bract primordium on its flank. Three leaf primordia are also visible; others have been removed.*

The floral phase therefore starts at floral initiation and continues until anthesis (*Fig. 1.1*). The timing of anthesis depends on the time of floral initiation and the duration of flower development, and should occur at a suitable time for good pod set and pod and seed growth.

FLORAL INITIATION STAGE

The floral initiation stage can be recognised when the shoot apex becomes conical (*Fig. 5.1*), rather than the gently rounded dome seen in the vegetative phase (*Fig. 3.3a*). A newly initiated flower primordium does not spread around the meristematic dome to form a crescent-shaped structure characteristic of a leaf primordium but forms a compact bulge, subtended by a ridge (the bract primordia).

The timing of floral initiation is determined largely by the final number of leaves which are formed on the main shoot and by the rate at which they are initiated (Chapter 3). The main factors affecting the final number of leaves are species, cultivar, and temperature and daylength, and hence location and sowing date (Chapter 3).

INFLORESCENCE FORMATION

The lupin inflorescence is a raceme, that is, the shoot apex is indeterminate; it does not become a single flower but produces many lateral flower primordia. Flowers are initiated more rapidly than leaves and in some cultivars about 40 are formed

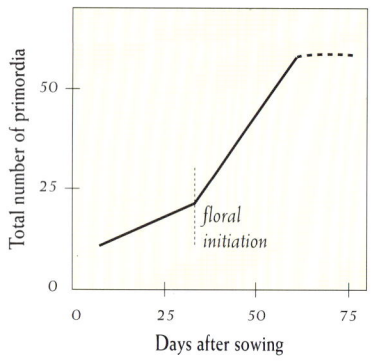

Fig. 5.2 *Total number of primordia (leaf and flower versus days after sowing) in Narrow-leafed Lupin. Note the increase in rate of primordium initiation after floral initiation. The dashed line indicates that the apex became abnormal and later degenerated.*

on the main shoot (*Fig. 5.2*). The number of flowers is highly variable, depending on plant size, vigour and competition from lateral branches.

In the early stages, individual flower primordia may be seen clearly on a dissected shoot apex (*Fig. 5.3a*), but later the hairy bract grows to cover the developing flower. As more flower primordia are initiated the overlapping bracts form a pear-shaped inflorescence (*Fig. 5.3b*).

When the inflorescence is about 10 mm long and all leaves have unfolded from the terminal bud, the inflorescence can be seen without dissection. At this stage the internode beneath the inflorescence (peduncle) and the inflorescence internodes start to grow and lengthen strongly. Later, the basal flowers on the inflorescence diverge from the inflorescence stem (rachis) and the bract falls off (*Fig. 5.3c*). At the same time the floral shoot apex degenerates.

DEVELOPMENT OF A FLOWER

The flower primordia initiated first remain the most advanced. All flower primordia have a similar course of development, although those towards the top of the inflorescence may degenerate before anthesis. The meristematic dome of each flower primordium initiates, more or less in sequence, the floral organs: sepals, petals, stamens and ovary. These organs develop at varying rates. The main features are shown in Table 5.1 and Fig. 5.4, which also shows the stages relative to shoot appearance. The flower primordium develops into a mature flower in about 40 days, the time taken depending on the environment and cultivar.

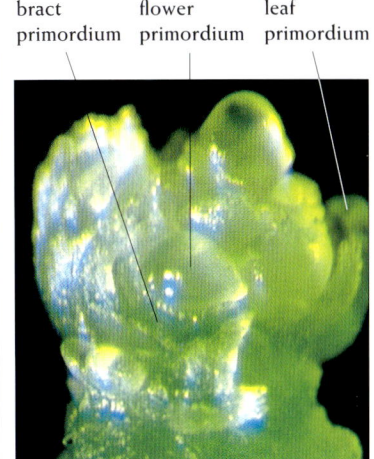

(a)

(b)

(c)

Fig. 5.3 *Three stages of inflorescence formation: (a) floral apex with leaf primordia, and several flower primordia; (b) young inflorescence with hairy bracts covering the developing flowers; (c) inflorescence with elongated peduncle in which the lowermost mature flowers are diverging from the rachis.*

Table 5.1 *Description of stages of flower development.*

DAYS AFTER FLORAL INITIATION	FLOWER STAGE	DESCRIPTION OF FLOWER DEVELOPMENT
0–10	Floral initiation	*Floral initiation* *Floral primordia present*
11–20	Organ primordia	*Bract well developed, hairy, covers flower* *Primordia of sepals, petals, stamens and ovary (margins not fused) visible* *Sepals and petals present as low ridges*
21–30	Anthers and ovary formed	*Sepals hairy, almost covering flower; petals beginning to grow; anther lobes developing; margins of ovary fused* *Sepals with long hairs; standard petal as long as the stamens*
31–40	Yellow anthers, stigma visible	*Keel petals half as long as stamens* *(All leaves unfolding; inflorescence visible)* *Keel petals covering stamens; stigma ridge developing* *Petals 1.5x stamens; wing petal envelopes keel; anthers yellow; papillae develop on stigma*
40+	Mature flower	*Standard petal completely enfolds other petals; margins of wing petals fuse (not strongly); staminal tube begins to grow; hairs grow on ovary.* *Bract falls; flower diverges from rachis; anthesis occurs*

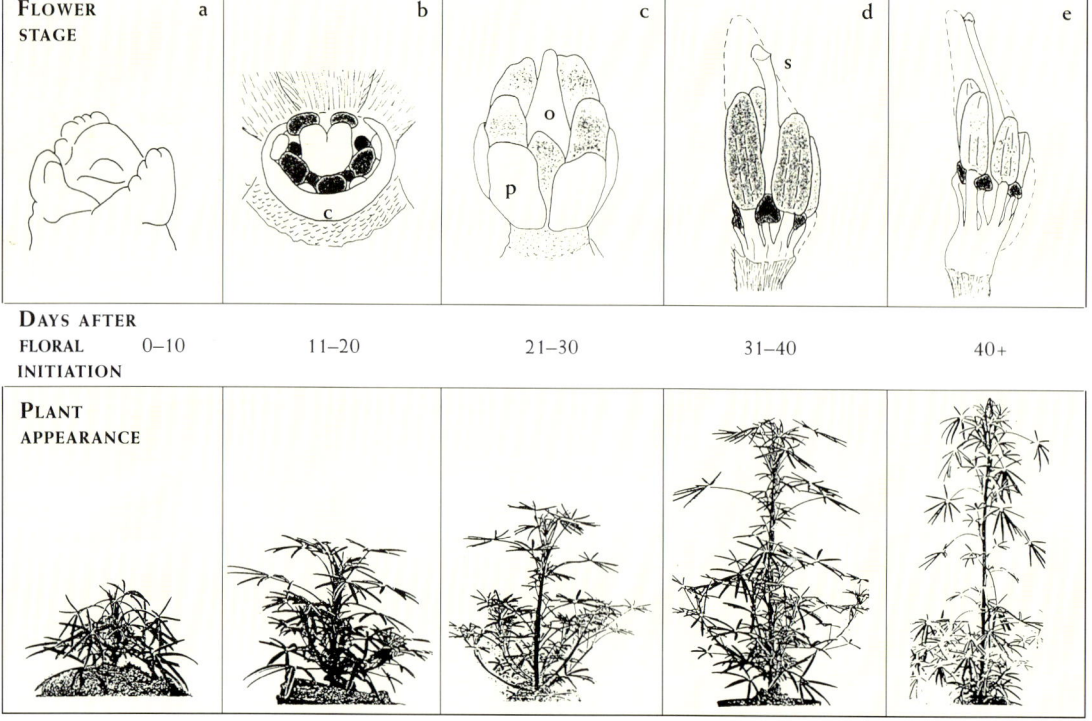

Fig. 5.4

Chapter 5: Inflorescence and Flower Development

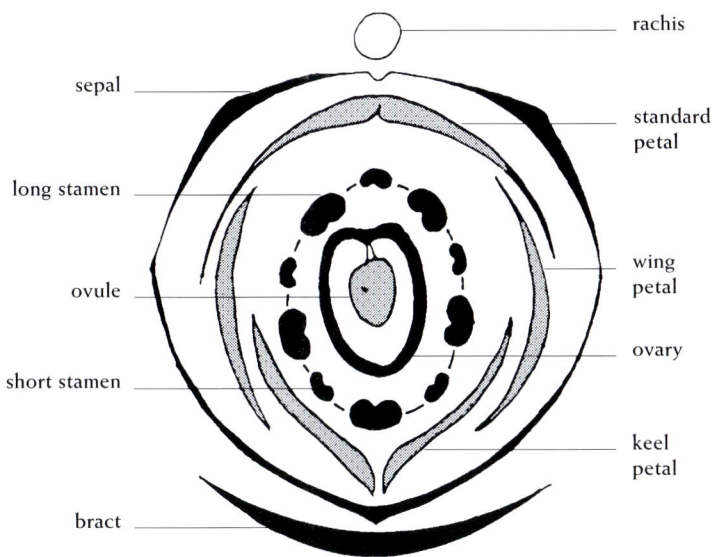

Fig. 5.5 *A mature flower. The standard petal is fully erect and anthesis has occurred. The pedicel, scar of bract, calyx and wing petal are shown.*

Fig. 5.6 *Floral diagram of lupin.*

STRUCTURE OF A MATURE FLOWER

The fused calyx (sepals) is very hairy and has a long lower (abaxial) pointed lobe and two shorter pointed lobes on the upper (adaxial) side (*Fig. 5.5*). The corolla has five free petals; the large uppermost petal is called the standard or vexillum. Two wing petals envelop the keel (*Fig. 5.6*). The two petals of the keel are fused along their margins except at the tip. The wing and keel petals have an interlocking mechanism which operates if the flower is 'tripped'. Tripping occurs when the

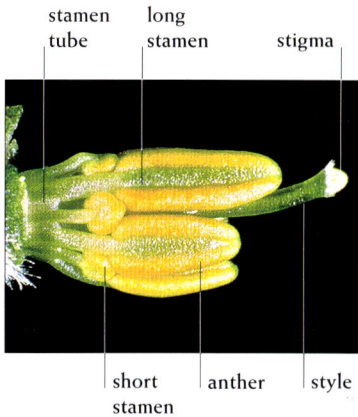

Fig. 5.7 *Flower of Narrow-leafed Lupin with calyx and petals removed, showing the stamen tube, short and long stamens, anthers, style and stigma.*

Fig. 5.4 *(Opposite) Diagram to illustrate Table 5.1. Upper panels: (a) floral initiation, the apex shown in side view; (b) organ initiation, the flower shown attached to the rachis. Each flower is covered by a hairy bract. In the diagram, the flower has the bract removed and is viewed from a high angle. In this figure, and in c, d and e, the anthers of the short stamens are darkly shaded, those of the long stamens lightly shaded; (c) anther and ovary formed, in a detached flower with the sepals removed and viewed from the front (abaxial) side; (d) yellow anthers and stigma visible, the flowers with sepals and all petals removed save one keel petal (shown as a dashed line), viewed from the side; (e) mature flower, the flower prepared as in (d).*
Lower panels: external appearance of plants at the flower development stage (not to scale).
Key to symbols: c, sepals (calyx); p, petals; o, ovary; s, style and stigma.

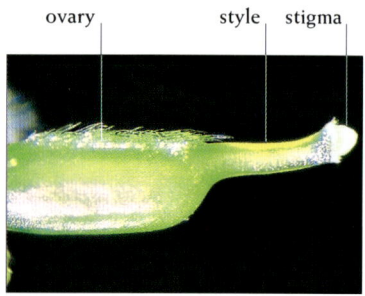

Fig. 5.8 *Ovary from flower in Fig. 5.7. It is densely clad in hairs and the style ends in the stigma, surrounded by hairs.*

weight of an insect landing on a flower depresses the wings and keel, exposing the style, stigma and stamens.

Ten stamens are joined together at the base to form a tube (the stamens are monadelphous). Stamens are of two types, one type having large anthers and longer filaments, the other having small round anthers and shorter filaments (*Fig. 5.7*). The short stamens are adjacent to the petals and the alternating long stamens are adjacent to the sepals.

The single ovary has a curving style terminated by a conical stigma below which is a fringe of hairs (*Fig. 5.8*).

—6—
ANTHESIS AND FLOWERING

Anthesis is a most important stage in the life cycle and can be recognised by changes in the appearance of the flower. In Narrow-leafed Lupin, anthesis occurs prior to flower opening so that self-pollination and fertilisation almost always occur. Cross-pollination, however, can occur in Albus Lupin and is common in Yellow Lupin. Flowers on the inflorescence open in sequence over a period of several days, but often only a few set pods. The remainder fall off after the petals wither.

ANTHESIS IS POSSIBLY the most important developmental stage in the life cycle of the plant. Pollination and fertilisation, which follow directly after anthesis, are the first events in the process leading to pod and seed growth and therefore yield is intimately related to the timing of anthesis.

The optimum time for anthesis is when the temperatures are neither too high nor too low for pollination and fertilisation and when the plant has sufficient photosynthetic capacity (leaf area) to meet the demands for nutrients from the growing pods and seeds. There should also be adequate time for the seeds to fill. In Mediterranean climates seed filling can be restricted by the onset of rising temperature and soil water deficits. In northern Europe seed filling may be curtailed by low temperatures and low light levels, and if seed filling is late in the season conditions may be difficult for crop ripening and harvesting. The time of anthesis in relation to the growing season is therefore very important in the adaptation of a cultivar to a

Table 6.1 *Stages of flower opening, senescence and pod set. The stage at which anthesis occurs is coloured in green.*

STAGE	DEFINITION	ANGLE TO RACHIS	
Pointed bud	Corolla above bract; standard tightly envelops wings and keel	0°	
Hooded bud	Standard enfolds less tightly; wings visible	0°	
Diverging standard petal	Standard unfolding from wings, not fully erect; bract falls off	35°	
Open flower	Standard usually fully erect	35°	
Blue corolla	Wing petals becoming blue	45°	
Senescent corolla	Petals withering	60°	
Incipient abscission and abscission	Abscission layer forming in pedicel; pedicel yellow, constricted at green/yellow junction then: Flower falls		
or **Pod set**	Pod clearly visible >8–10mm; withered remnants of petals at base	60°	

The effect of species, cultivar, temperature, daylength and vernalisation on the time of anthesis can be measured and summarised in the form of mathematical functions (Chapter 10). This allows prediction of the effect of variation in weather on the adaptation of cultivars to different areas.

Flower Opening

After an inflorescence emerges from the enclosing leaves and flowers become visible, emerging from within their covering bracts, a flower goes through a number of stages (Table 6.1). These are defined by flower shape, colour and angle to the inflorescence rachis. In the pointed bud stage the margins of the wing petals are weakly fused but separate as the flower unfolds. Anthesis coincides with the stage when the flower begins to diverge from the stalk and the standard petal begins

Fig. 6.1 *Flower just after anthesis. The flower has diverged from the rachis and the standard petal is erect.*

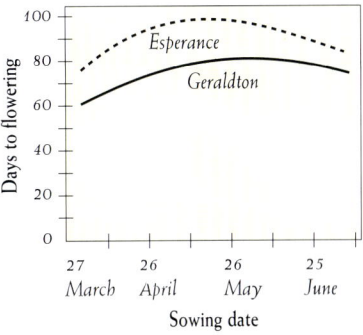

Fig. 6.2 *Expected time to flowering of Narrow-leafed Lupin in Western Australia at Geraldton (north-western end of the lupin growing area) and Esperance (south-eastern end of the lupin growing area) for various times of sowing.*

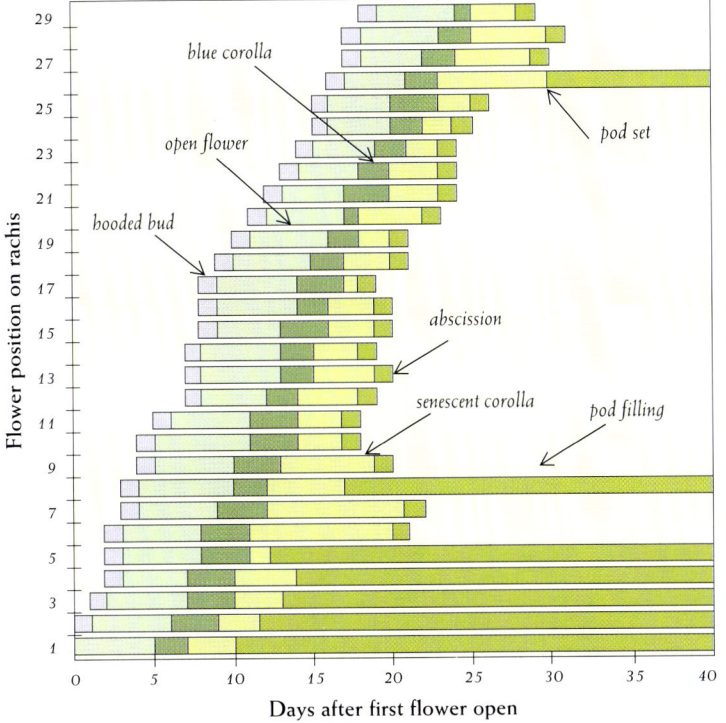

Fig. 6.3 *Diagram showing the timing of stages of flower opening and pod set on a typical main shoot inflorescence.*

Fig. 6.4 *A flower just before anthesis. It has been cut open to show the long and short stamens and the stigma protruding beyond the stamens into the tip of the keel petals.*

to unfold and become erect (*Fig. 6.1*). This stage, 'diverging standard petal', is transient and takes less than a day.

The most basal flower on the inflorescence is the first to reach anthesis and open flower stage (*Table 6.1*). In Western Australia, Narrow-leafed Lupin sown late in April will reach anthesis 75–100 days after sowing, depending on its location within the State (*Fig. 6.2*) (earliest flowering occurs in northern areas where temperatures are higher and days longer during winter). Flower opening then proceeds up the inflorescence and one to three flowers open every day (*Fig. 6.3*). Typically, about 30 flowers may open on a main shoot inflorescence, taking about 20 days. Branches bear fewer flowers and the duration is shorter, but otherwise the process is similar.

ANTHESIS AND POLLINATION

As a flower approaches anthesis the stigma, which is well above the long stamens, becomes sticky (*Fig. 6.4*). At anthesis the filaments elongate and the anthers of the long stamens shed their pollen inwards, forming a plug or mass of pollen. The filaments of the long stamens then curl up and play no further part (*Fig. 6.5*). The filaments of the short stamens continue to grow and also thicken, pushing a plug of pollen into the tip of the keel petals where it surrounds the stigma. Later, the small anthers shed their pollen. Pollen grains forced into contact with the stigma germinate and downward growth of the pollen tubes inside the style transfers the male gametes to the ovary and fertilisation ensues. As far as is known there is no difference in the effectiveness of pollen from long and short stamens.

SELF- OR CROSS-POLLINATION

Lupin pollen comes into contact with the stigma without the intervention of any pollinating agent. In contrast to some genera of *Fabaceae*, pollen transfer in the cultivated lupins does not depend on the flower being tripped. (Flower tripping occurs when a sufficiently large insect lands on the flower and forces down the interlocking wings and keel, so that the style and stigma protrude from between the wings. During tripping, the stigma may be pollinated with either self pollen or pollen carried on the insect from another flower.) In some species of lupin and other genera of *Fabaceae*, tripping is necessary to bring pollen in contact with the stigma and to prepare the stigma to ensure pollen germination. Bees are frequent visitors to lupin flowers to collect pollen, although by the time that the flower of Narrow-leafed Lupin is sufficiently open to attract them the stigma is already smothered with self pollen. Lupin pollen is too large and sticky to be carried on the wind.

In Narrow-leafed Lupin, flowers are generally self-pollinated and visits by insects do not increase the frequency of pod set or yield. The amount of cross pollination is very low which aids the maintenance of pure line cultivars.

Albus Lupin has a similar pollination mechanism to Narrow-leafed Lupin, and is also visited by bees. Cross-pollination is more frequent than in Narrow-leafed Lupin and levels of 10 per cent have been reported. Yellow Lupin is the most promiscuous of the cultivated lupins and 40 per cent of out-crossing is possible.

POD SET AND ABORTION

Only a proportion of flowers on an inflorescence form mature pods; some drop off after the stage of petal senescence, about 10 days after anthesis and despite having been fertilised. In modern varieties of Narrow-leafed Lupin, pods are set from a higher proportion of flowers than in older varieties. Nevertheless, it is common for up to 90 per cent of flowers to be shed and therefore not set pods.

A flower has set a pod when first the style and later the tip of the pod protrude beyond the withered petals. In those flowers that do not form pods, the first sign of their impending fate is a colour change of the flower stalk (pedicel) from green to yellow. This is associated with the formation of an abscission layer at the base of the pedicel (*Fig. 6.6*). Once the abscission layer is fully formed the flower falls off, whereas in those flowers in which the layer has not formed the ovary starts to grow rapidly into a pod (*Fig. 6.7*).

Fig. 6.5 *A flower, cut open to show anthesis. The anthers of the long stamens have dehisced and the filaments have curled up. The short stamens have elongated and their anthers form a 'piston' which forces a plug of pollen into the tip of the keel and onto the stigma.*

Fig. 6.6 *Longitudinal section of a pedicel and bract, showing where the abscission layer forms in the pedicel.*

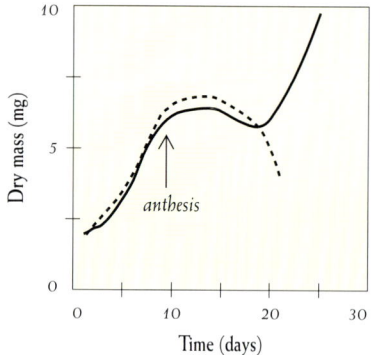

Fig. 6.7 *Growth of the ovary of Narrow-leafed Lupin at a position on the raceme with a high frequency of pod set (——) and at a position where all flowers abscise (----).*

Flowers are shed from the main shoot when the lateral branches begin to grow vigorously and when some pods have already set. In experiments where the branches have been cut off at an early stage of development, flower shedding is reduced. These observations suggest that flower shedding involves competition for resources between the main shoot and the branches. This is consistent with the greater pod set on the main shoot of lupins with restricted branching (Chapter 4).

Many chemical treatments have been tried in an effort to improve pod set. Applying a hormone called cytokinin to flowers has been the only notable success, but it must be applied to each flower pedicel as the flower approaches anthesis. Increased pod set does not, however, increase yield as some of the pods fail to fill and other yield components adjust to compensate for the presence of more pods.

Pods may abort, albeit at a lower frequency than for flower shedding, after pod set. When they abort well after pod set the pods may remain attached to the plant but contain only unviable seeds. Seeds can also abort, most frequently during the early stages of their growth and before rapid seed filling starts. The seeds most likely to abort are those at either end of the pod.

—7—
POD AND SEED DEVELOPMENT

The final phase of the life cycle is pod and seed development. The pod grows in size and, as it approaches maturity and dries out, changes in colour from green through khaki to light reddish-brown. In the pod there are up to five seeds. Seed growth initially lags behind growth of the pod walls but continues for longer. At maturity, each seed contains a well developed plant with root, cotyledons, young leaves and a shoot apex. The moisture content of the seed changes as it matures; at maximum dry mass it is about 62 per cent and declines to about 12 per cent at harvesting. The seed changes from green to pale fawn at maturity and the cotyledons from green to golden. Growing conditions during pod and seed development affect not only crop yield but also seed quality. Colour changes of the pod and seed are useful indicators of physiological stages on which to base management decisions such as spraying trace elements and herbicides (e.g. crop topping), swathing and harvesting.

THE EARLY STAGES of pod and seed development and growth coincide with leaf and branch growth elsewhere in the plant, so strong competition arises between these organs. The success in filling an individual pod is a complex relationship involving the competitive relationships between organs in the canopy, the time when filling starts and environmental conditions.

A healthy plant and good canopy at the beginning of pod filling are important for high yield potential. The conditions during pod and seed filling largely determine to what extent this yield potential is realised. In Western Australia, the growing season is ended by dry weather, and the timing and rate of development of the consequent drought have a large influence on the ability of the plant to realise its yield. Depending on their severity and the stage of development of the seeds and pods, high temperatures, frost and waterlogging can also reduce pod filling.

Canopy architecture can also have a strong influence on the likelihood of pods setting and filling. For example, restricting branching by removing branches or by using genotypes with restricted branching increases pod set on the main shoot. Conversely, in well grown, freely branching crops, pods on heavily shaded low branches may be out-competed by those on higher branches and abort.

POD GROWTH

After pollination and fertilisation, the ovary of some flowers on the inflorescence develops into a pod. The probability of a pod being set decreases up the inflorescence. After pod set, at which stage the pod is 8–10 mm long, the probability of abortion is much reduced. Pods that abort before or shortly after pod set generally fall off; those that abort later may remain attached to the plant but bear no seed.

Fig. 7.1 *Stages of pod growth of Narrow-leafed Lupin: (a) young pod soon after pod set; (b) green pod with position of seeds visible; (c) maturing pod becoming khaki-coloured; (d) mature dry pod; (e) completely ripe pod showing eventual opening.*

Pods (legumes) begin as a grass-green colour and soon grow bigger than the enclosing senescent petals which generally fall off as the pods grow (Fig. 7.1a). The sepals persist, as does the stamen tube, but this is soon split by the growth of the pod. As the seeds grow, their positions can be seen as bulges in the pod wall (Fig. 7.1b).

The early stages of pod growth are marked by rapid lengthening and then thickening of the walls (valves). In the increasingly thick and succulent pod, seeds become enclosed in compartments formed by the growth of septa (cross walls) between the seeds (Fig. 7.2). As the seeds grow they force the pod walls apart, rupturing the septa. The seeds are attached alternately to opposite walls by short stalks (funicles) (Fig. 7.3).

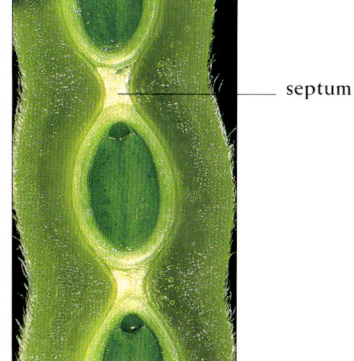

Fig. 7.2. *Photograph of part of a pod of Narrow-leafed Lupin, cut longitudinally to show the septa formed between seeds and the succulent pod walls.*

The thick pod walls serve as a reservoir of nutrients for the seeds, and pods reach maximum weight before the seeds are even half filled. After the pod reaches maximum weight, the walls lose much of their contents to the seeds, so that the seeds continue to enlarge while the pod walls shrink. At this stage the walls change from khaki to a pale reddish-brown (Figs 7.1c, 7.4). At maturity the hard, leathery, pale straw-coloured walls represent a comparatively high 35 per cent of the weight of the pod.

Undomesticated lupins have pods that shatter at maturity, that is, as the pod walls dry out they spring apart, twisting as they do so, and scattering the seeds. This character was removed by breeding and modern cultivars do not readily shatter. Nevertheless, the pod becomes brittle and will eventually open spontaneously, so lupin crops should be harvested as soon as feasible.

SEED AND EMBRYO GROWTH

The embryo is formed by the fusion of egg and sperm cells and during its growth is nourished by the endosperm. At the beginning of pod filling the seeds grow slowly compared with the pod. By the time the seed begins to grow appreciably the embryo is clearly visible, immersed in jelly-like and liquid endosperm (Fig.7.5a) which, together with temporary reserves

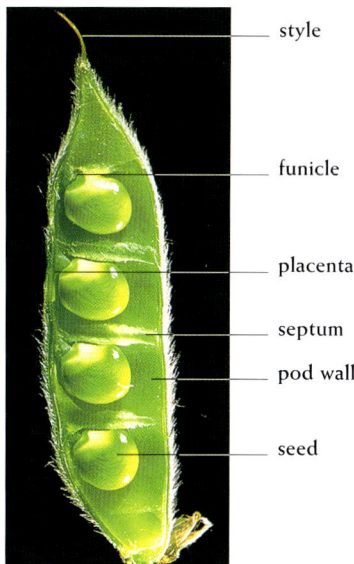

Fig. 7.3. *Photograph of a pod of Narrow-leafed Lupin. One wall has been removed to expose the seeds. The left hand margin of the pod is uppermost.*

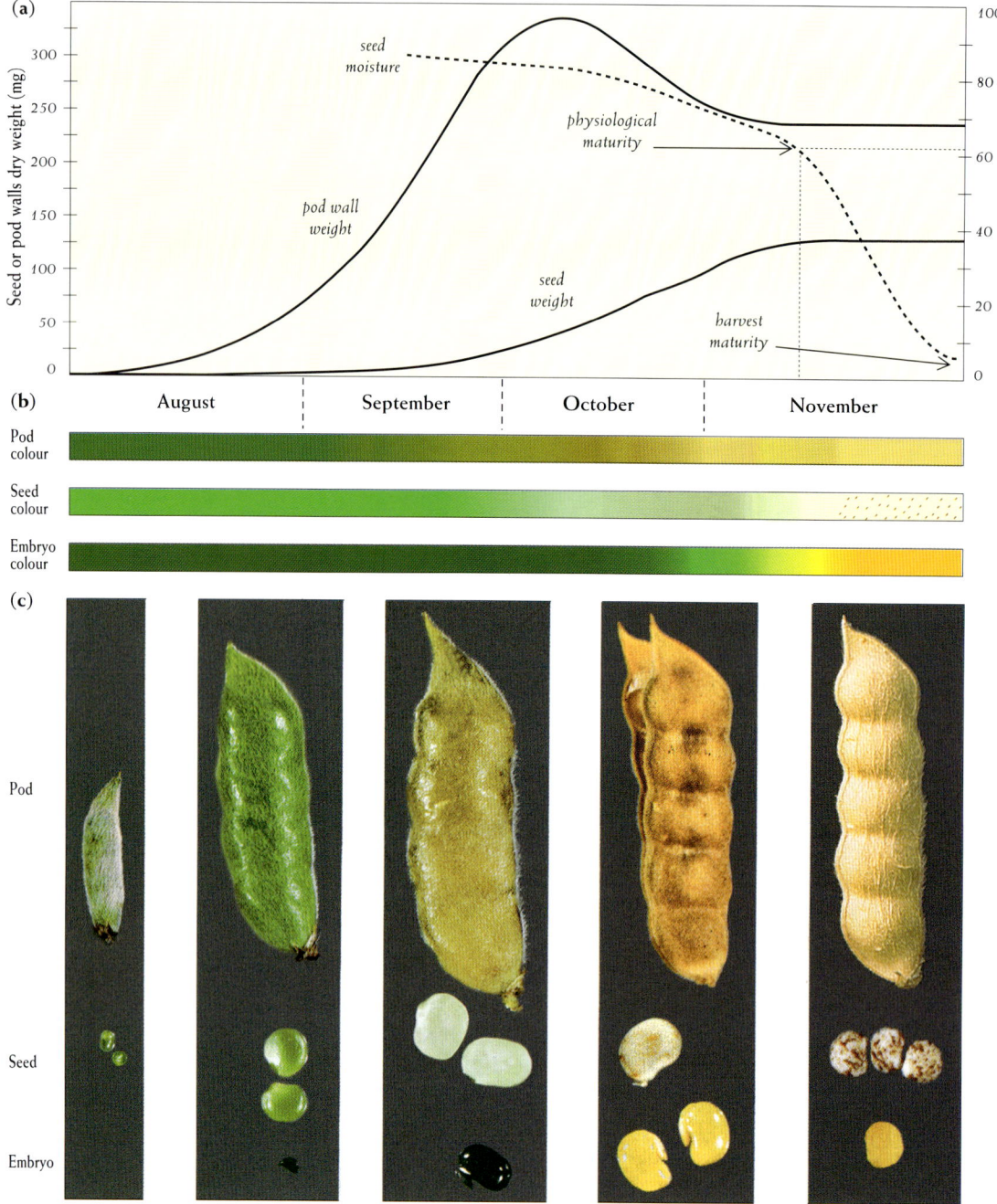

Fig. 7.4 Pod and seed development of Narrow-leafed Lupin. Dates are typical for a crop grown in Western Australia

(a) Dry weight of pod walls and individual seeds, and moisture content of the seeds.

(b) Coloured bars show changes in the colours of the pod, seed and embryo.

(c) Pictures show the pod and representative seeds, and embryos from those pods, at various stages in their development.

stored in the seed coat, nourishes the developing embryo. The embryo develops two cotyledons, a radicle, a hypocotyl, a shoot apex with five or six leaf primordia, and a meristematic dome. As the seed matures the shoot apex is subjected to considerable pressure from the growing cotyledons and the tissues become brittle. Leaf primordia and the shoot apex become difficult to examine and it is not possible to count leaf primordia.

The water content of seeds declines slowly until seeds reach physiological maturity, at which time they contain about 62 per cent water (Fig. 7.4). At physiological maturity the seed has reached its maximum dry weight, seed filling is complete and the seed is fully capable of growing into a new plant. After physiological maturity the seed loses moisture rapidly (Fig. 7.4) and shrinks considerably until harvest, when it contains about 12 per cent water and occupies only about half of the volume inside the pods.

For much of their growth seeds are bright green, but when they are about half grown the coat colour changes to pale bluish-grey and later to a pale fawn (Figs 7.4, 7.6). Meanwhile, the cotyledons become golden at physiological maturity (Figs 7.4, 7.5c). In cultivars with speckled seed coats, the speckling develops soon after the seed reaches physiological maturity (Figs 7.4, 7.6d).

(a)

(b)

(c)

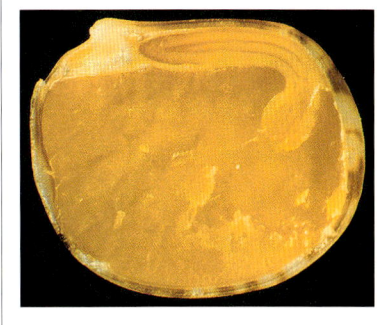

(d)

Fig. 7.5 *Stages of embryo development in Narrow-leafed Lupin. The seed and pod were cut longitudinally, removing one cotyledon to expose the shoot apex (except in a); (a) young embryo, surrounded by jelly-like endosperm; (b) later stage, with the radicle in the radicle pocket and the shoot apex and radicle vascular tissue visible and the seed coat is swollen; (c) embryo at physiological maturity, with yellowing cotyledons, the shoot apex and radicle are well developed (and radicle is in radicle pocket), the endosperm is exhausted and the seed coat has shrunk; (d) mature seed, where the golden embryo has become brittle.*

(a)

(b)

(c)

(d)

Fig. 7.6 *Stages in seed development in Narrow-leafed Lupin; (a) young, green seed showing developing septum between seeds; (b) almost fully grown seed, septa have split; (c) pale greyish-blue seed coat, embryo fully developed; (d) mature seed; note shrinkage.*

SEED ABORTION

In some pods not all seeds grow to maturity. Missing seeds in the pod may be caused by non-fertilisation or abortion at varying times during growth, due to environmental stresses such as high temperature, frost or inadequate nourishment. There may be no sign of a seed other than an empty space in the pod, or the undeveloped seed may persist. Seed abortion is most likely at positions at either end of the pod.

POD AND SEED GROWTH ON THE BRANCHES

Growth of pods on the branches follows a similar course to that described for the main shoot. They go through similar colour changes and the relation of pod colour to moisture content is the same. Seed and embryo development also follow patterns similar to those on the main shoot.

Pod and seed growth on the branches start later than on the main shoot, but they grow faster on the branches, so that pods and then seeds reach maximum weights at similar times throughout the plant. Similarly, seeds reach harvest maturity at much the same time, in marked contrast with older cultivars and wild lupins where seeds ripen sequentially up the plant.

—8—
THE ROOT SYSTEM

Root growth begins with germination, when the radicle protrudes through the seed coat and forms the tap root. Root elongation results from division and elongation of cells at the root tip, which is forced through the soil, and is protected from abrasion by the root cap. Root hairs appear about 1 cm behind the root tip. First-order lateral roots begin to appear about 10 cm from the tip and in turn form second- and third-order laterals. Nodules develop mainly on the tap root. Soil conditions and management affect the rate of growth and distribution of roots and thus the capacity of the plants to extract nutrients and water from both the nutrient-rich surface soil and from deep in the profile.

ROOTS PROVIDE PHYSICAL support for the shoot and the nitrogen fixing nodules, and explore the soil for mineral nutrients and water. A properly developed root system is important to meet these requirements adequately. Mineral nutrients are usually concentrated in surface layers of soil but also leach downwards, necessitating good elongation and ramification of roots through the soil. Since nitrogen leaches readily, especially in sandy soils, good nodulation confers an adaptive advantage for plants grown on these soils. Water availability near the surface may fluctuate throughout the growing season and in Western Australia declines rapidly toward the end of the season, so root activity deep in the soil profile also becomes important.

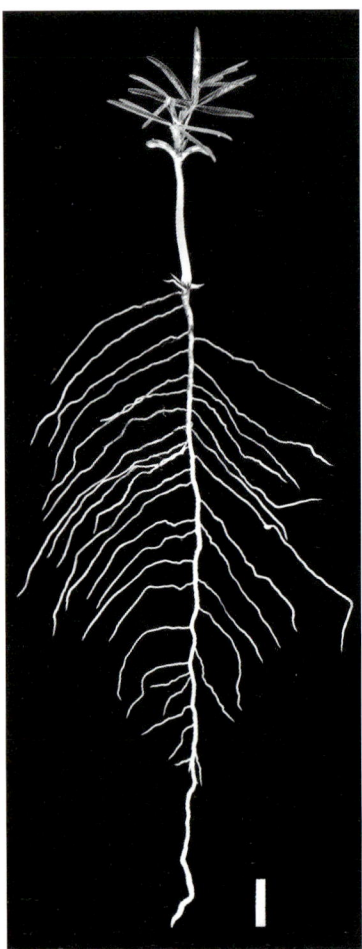

Fig. 8.1 *A three-week-old Narrow-leafed Lupin plant showing a well developed root system. The first-order lateral roots begin about 10 cm from the root tip. Bar represents 2 cm.*

Many factors influence the distribution and intensity of development of the root system, particularly soil management which affects compaction, waterlogging and fertiliser distribution. As with the shoots, roots have an organised pattern of development, and knowledge of this can be helpful in deciding on which soils to grow lupins and the management practices to be adopted.

Development and Morphology

When the seed germinates, the radicle appears first (Chapter 2) and grows to form a tap root about 1.5–2 mm in diameter.

Root elongation leads to the delicate root apex being forced through the soil. A cap over the apex protects it from soil abrasion. The root cap is a thin, thimble-shaped mass of loose, slimy cells. The root cap cells, sometimes called 'root border cells' are constantly being produced by cell division at the root apex and are sloughed off as the root tip grows through the soil. Apart from lubricating the passage of the root tip, the root cap cells probably contribute much of the mucilage around the root (see below). Root elongation results from division and elongation of cells within 1–1.5 cm of the apex.

Behind the root elongation zone, root hairs appear on the root surface. These are prolific (about 500 per sq. mm), fine (less than 0.01 mm in diameter) and short (less than 0.5 mm), just visible with the naked eye. They greatly increase the root surface area, aiding nutrient and water uptake.

If young roots are gently removed from soil, a mucilaginous sheath of adhering soil ('mucigel') is visible. The mucigel is composed of plant mucilages, rhizosphere micro-organisms and their products, and soil material. It may have several functions: a selective niche for rhizosphere microbes; to protect the root from desiccation and/or nutrient toxicities; and to promote contact between the root and soil to assist transfer of nutrients to the root.

First-order lateral roots appear 10–14 days after sowing and 10 cm behind the root apex (*Fig. 8.1*), but the distance can be considerably shorter if root extension is restricted, for example by soil compaction. About two first-order laterals appear for every centimetre of tap root and they have a developmental pattern similar to the tap root. However, whereas the tap root grows mainly downwards, the first-order laterals tend to grow in a more horizontal direction for several centimetres before turning downwards. The first-order laterals are thinner than the tap root, and over time they produce even thinner laterals (second-order laterals) which in turn form third-order laterals. In these higher order laterals the direction of growth is more or less random.

Adventitious roots may appear on the hypocotyl below the soil surface about 5 weeks after sowing (*Fig.* 8.2). These roots can be numerous, have little or no branching and are most pronounced when the soil stays moist or after waterlogging.

Some lupin species such as *L. cosentinii*, *L. pilosus* and Albus Lupin, but not Narrow-leafed or Yellow Lupin, can also form cluster (or proteoid) roots which enhance nutrient uptake.

Fig. 8.2 *Ten-week-old Narrow-leafed Lupin plant showing healthy development of nodules on the tap root and adventitious roots arising from the hypocotyl.*

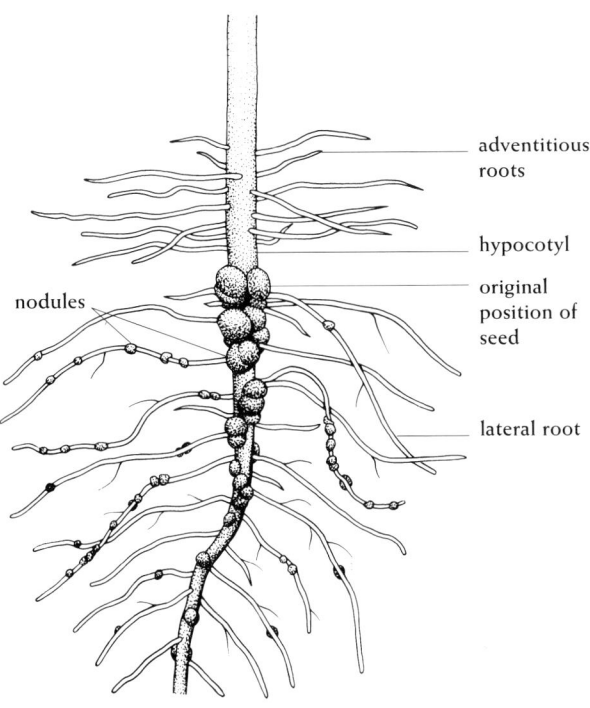

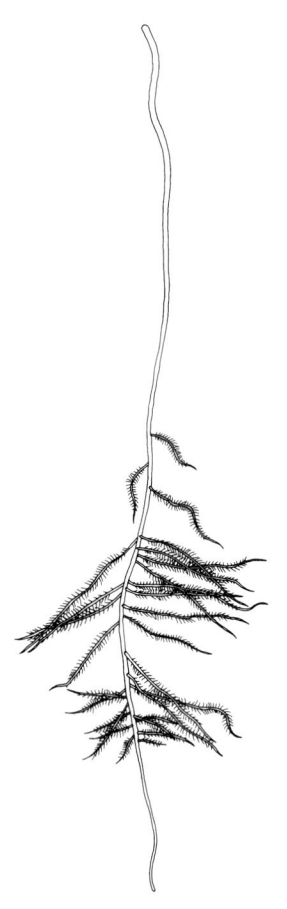

Fig. 8.3 *Lateral root of Albus Lupin showing development of cluster (or proteoid) roots.*

These are a dense proliferation of short roots covered in a thick mat of root hairs (*Fig.* 8.3). Formation of cluster roots is reduced when phosphate is added to the soil.

Only low levels of infection with mycorrhizal fungi occurs in lupins.

Young roots are not pigmented but become brown as they age. The diameter of older parts of the root system increases progressively, especially in the tap root which becomes woody. The tap root and first-order laterals predominate in the root system of cultivated Narrow-leafed Lupin, which adapts the plant to extracting water from deep in the soil profile late in the season. First-order lateral roots tend to be longer and more prolific in White Lupin.

Nodules

Lupins form symbiotic relationships with nitrogen-fixing bacteria (rhizobia; *Bradyrhizobium lupini*). Following infection with these bacteria, lupinoid nodules form, which are ruggedly rounded outgrowths from the root. Healthy nodules are bright pink to red inside due to leghaemoglobin which is involved in the fixation of gaseous nitrogen. Nodules are found mainly on the top 5–10 cm of the tap root where, in a healthy crop, they eventually girdle the root (*Fig.* 8.2). Nodules are sparse on the lateral roots unless the tap root is lost.

To establish the symbiosis, roots must come into contact with, and be infected by, the rhizobium. When sowing the first lupin crop in a paddock (field), seeds should be inoculated with rhizobia to ensure nodulation (seeds are inoculated by mixing them with a bacterial inoculum and a gum solution to bind the inoculum to the seed). The bacteria persist in the soil for about ten years after a well nodulated lupin crop.

Nodulation is affected by temperature and therefore time of sowing, but nodules are usually visible on roots within 3–4 weeks of sowing and start fixing nitrogen within 5 weeks of sowing. Leaf photosynthate is the source of energy for nodule growth and respiration, and for nitrogen fixation and assimilation. Nitrogen fixation is therefore closely related to photosynthesis and increases as the leaf canopy develops. Nitrogen fixation is greatest at or close to flowering, then declines and terminates shortly after pronounced leaf drop. The nodules consume up to one-fifth of the plant's net photosynthate; of this, half is combined with fixed nitrogen and re-exported to the plant, 40 per cent is lost in respiration and 10 per cent is incorporated as nodule dry matter.

Growth

Length

Many factors influence the growth, morphology and development of roots, including temperature and structure of the soil, moisture and nutrient availability, and soil compaction and waterlogging. Soils at field capacity and temperatures of 18–22°C are most favourable to root growth. Although soil compaction restricts root growth, the comparatively thick roots of lupins appear able to push their way through compact layers better than roots of most crops. However, limitations to oxygen diffusion across thick roots during waterlogging might make the crop more sensitive to waterlogging.

The tap root of Narrow-leafed Lupin can grow at 2.5 cm per day, reaching 70 cm after a month and often exceeding 2 m in depth by maturity in deep, coarse-textured soils (*Fig. 8.4*). Lupins are therefore able to extract water from comparatively deep soil profiles. Roots are most abundant in surface soil layers, and at flowering almost half of the root length is within the top 20 cm of soil. Nonetheless, other crops such as wheat have a greater proportion of their roots in the surface soil. Root distribution also responds to nutrient distribution and moisture availability, roots growing preferentially in moist layers where fertiliser has been applied, such as with deep banded phosphate.

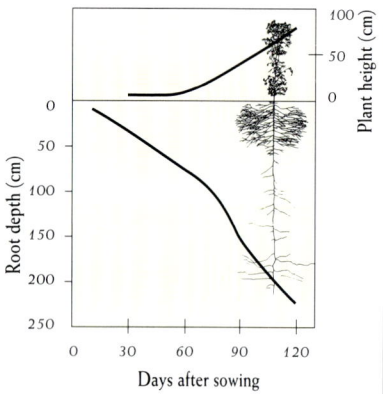

Fig. 8.4 *Root extension of Narrow-leafed Lupin down a deep loamy sand profile in Western Australia.*

Total root length of Narrow-leafed Lupin is low compared with many crops, such as wheat, but higher than in Yellow Lupin. Root length reaches a maximum close to the time of pronounced leaf drop. In Western Australia, up to two kilometres of root length per square metre of ground may develop under Narrow-leafed Lupin crops. Shortly after flowering ends, roots in surface layers decay but those at depth continue to grow.

Weight

At germination, the radicle is an insignificant proportion of seed weight. In Narrow-leafed Lupin the root then grows considerably faster than the shoot, and six weeks after sowing can account for over half the plant's weight. In Yellow Lupin, roots are a considerably smaller proportion of plant weight. The proportion of plant weight in roots of Narrow-leafed Lupin declines to between one-fifth and one-half at flowering and then decreases rapidly; at maturity roots account for only about 3 per cent of plant weight.

The root system (excluding nodules) consumes about one third of the net photosynthate, so that factors which reduce leaf area will reduce root growth. Under difficult soil conditions such as dry, infertile or compact soil, the plant tends to allocate a greater proportion of its resources to root growth. For example, although roots can not grow into completely dry soil, seedlings growing in soil with only a little moisture have comparatively deep roots when there is still only little shoot growth. This promotes exploration of the soil for moisture.

—9—
LUPIN DEVELOPMENT SCALE

Lupins develop in a definite sequence, progressing through several clearly distinguishable development phases between sowing and maturity. Crop management and assessment depend on reliable, timely identification and communication of stages during the plant's development. The Lupin Development Scale outlines discrete stages of development. The scale is based on the Zadoks scale developed for cereals. Each stage is identified with a brief description and is given a unique numerical code to aid rapid recording. Diagrams are also provided to aid identification of key stages.

DEVELOPMENT OR GROWTH stage scales aid the communication about development stages and are therefore a valuable tool to researchers, agronomists, advisers and growers. Use of a clearly defined, illustrated standard scale for assessing development will enable recommendations for the optimum timing of agrochemical applications to be established precisely and then communicated. Development stages can also be used as a basis for comparing crops and thus identifying those that might have growth problems.

Development stages distinguished in a scale should be easily recognisable in the field, with little need for specialised knowledge or equipment. Descriptions of defined stages should be objective so that there is little or no difference in recognition of stages between users. A scale which recognises developmental stages similar to those recognised in scales for other crops would also help with comparisons between crops.

The scale described below for lupin follows conventions similar to one for cereals (developed by Zadoks, Chang and Konzak) which has already been adapted to other crops such as soya bean and faba bean. The scale is designed for use with individual plants and focuses on the stages of development of the main shoot. Various development stages are illustrated to aid their identification.

The sequential development in lupin plants, both within a shoot and between shoots, complicates the construction and use of development scales. In the scale below, flower, pod and seed development are assessed for the lowest (most advanced) node of the main shoot inflorescence. Focusing on the lowest node of the main shoot allows precision in describing development and will satisfy most requirements of a scale. Stages of development of the branches are generally similar to those of the main shoot, hence the same scale may be applied to the branches. Furthermore, the developmental gradients within inflorescences and between shoots become comparatively small toward the end of the life cycle, so as maturity approaches, descriptions made on the main shoot using this scale become more and more applicable to the whole plant.

If more general descriptions of inflorescences are needed, users can construct their own descriptions. Examples of how this might be done, focusing on the important stages of open flower, physiological and harvest maturity, are presented within the development stage definitions and allow for the assessment of the proportion of individual inflorescences that have reached those particular stages. Suggestions are made for extending these descriptions to the whole plant.

The scale describes 5 principal development stages (*Table 9.1*). Such broad descriptions are adequate for some purposes but for others greater detail is necessary; this is provided by sub-dividing each principal stage into secondary stages (*Table 9.2*).

Each development stage is identified numerically, and unlike the Zadoks scale a point separates the principal development stage from the secondary development stage, for example, 1.2 is the second stage of the first principal development stage. The number of secondary stages distinguished varies between principal development stages. The secondary stages are numbered in chronological order, but some numbers have not been allocated, e.g. 0.2, and these can be used subjectively to indicate an advanced state of the preceding stage. There is often overlap between stages so that a plant's development stage may be described by numerals for secondary development stage categories within two or even more principal development stages; in that case the user may choose to restrict descriptions to only the highest numeral or the stage of greatest interest.

Although numerical codes are applied to each stage, they are intended only for field recording and not for use in numerical analysis. Further, references to development stages should, where possible, make full use of the descriptive phrases of the code. Such descriptions may be supplemented with more precise records, such as from apical dissections, leaf area indices, or numbers of branches.

To assess the development stage of a crop a random sample of representative plants must be described (see Chapter 11). Techniques for identifying development features of a crop are described in Chapter 11. Development of the crop can then be classified by the highest numerical stage reached by a majority of plants.

Table 9.1. *The principal development stages for lupin.*

0	Germination
1	Leaf emergence
2	Stem elongation
3	Flowering
4	Pod ripening
5	Seed ripening

DEVELOPMENT STAGE DEFINITIONS

GERMINATION (CHAPTER 2)
The radicle (which becomes the tap root) is the first part of the seedling to protrude through the seed coat, followed by the hypocotyl. When the radicle is 5 mm long the seed has germinated (*Fig. 9.1*). The radicle and hypocotyl can be distinguished from each other by a slight colour change, especially if held up to the light. The hypocotyl hook (*Fig. 2.2*)

Table 9.2 *Descriptions of the principal and secondary development stages for lupin.*

0	GERMINATION & SEEDLING EMERGENCE
0.0	Dry seed
0.1	Start of imbibition (water absorption)
0.2	....
0.3	Radicle (root) protruding through the testa (seed coat)
0.4	....
0.5	Radicle 5mm long (germination) (Fig. 9.1)
0.6	....
0.7	Hypocotyl protruding through the seed coat
0.8	....
0.9	Part of seedling protruding above the soil (emergence) (Fig. 9.2)

1	LEAF EMERGENCE
1.0	First pair of leaves protruding beyond upright cotyledons (Fig. 9.3)
1.1	1 leaf emerged from bud (coincident with 1.2)
1.2	2 leaves emerged from bud
1.3	3 leaves emerged from bud (Coincident with 1.4)
1.4	4 leaves emerged from bud
1.5	5 leaves emerged from bud
1.6	6 leaves emerged from bud
1.7	7 leaves emerged from bud (Fig. 9.4)
:	....
1.10	10 leaves emerged from bud
1.11	11 leaves emerged from bud
:	....
1.n	n leaves emerged from bud

2	STEM ELONGATION
2.0	....
2.1	Little separation between bases of leaves (Fig. 9.5)
2.2	....
2.3	Bases of some basal leaves clearly separated
2.4	....
2.5	Bases of several leaves clearly separated from each other (Fig. 9.6)
2.6	....
2.7	Inflorescence bud clearly visible (bud) (Fig. 9.7)
2.8	....
2.9	Inflorescence bud clearly separated (by peduncle) from base of highest leaf (Fig. 9.8)

3	FLOWERING
3.0	Bracts completely hiding corolla
3.1	Pointed bud stage
3.2	Hooded bud stage
3.3	Diverging standard petal stage (anthesis)
3.4	Open flower stage (Fig. 9.9)
3.5	Coloured corolla stage
3.6	....
3.7	Senescent corolla stage
3.8	Floret abscised
3.9	Pod set (Fig. 9.10)

4	POD RIPENING
4.0	Young, green pod. No septa between seeds, seeds abutting (Fig. 9.11)
4.1	Young green pod. No septa between seeds, seeds separated
4.2	Green pod, septa between seeds, slight bulging of walls, seeds filling 50 per cent of space between septa (Fig. 9.12)
4.3	Seeds filling 75% of space between septa
4.4	Green pod, clear seed bulges in pod walls, seeds filling all space between septa
4.5	Green pod, septa split (Fig. 9.13)
4.6	....
4.7	Pod turning khaki-coloured
4.8	....
4.9	Pod pale reddish-brown and wrinkled

5	SEED RIPENING
5.0	Seed small, dark green, with watery contents
5.1	Seed medium size, dark green, with watery contents
5.2	Seed large, dark green, with watery contents
5.3	Seed large, green, little watery contents
5.4	Seed large and soft, light green coat, no watery contents, green cotyledons
5.5	Seed large, light green to pale greyish-blue coat, green cotyledons
5.6	Seeds large and soft, light green to pale greyish-blue coat, green to yellow cotyledons
5.7	Seed large and soft, pale fawn coat, yellow to golden orange cotyledons (physiological maturity) (Fig. 9.14)
5.8	Seed hard but dentable, mottling (in some cultivars) of pale fawn coat,
5.9	Seeds hard and ripe for harvest (Fig. 9.15)

is the first part of the seedling to appear through the soil surface; as soon as this take place, the seedling has emerged (*Fig.* 2.3, *Fig.* 9.2). For a population (crop) of plants, emergence is defined as the time when 50 per cent of the eventual total number of seedlings have emerged (Chapter 11). When the mean daily temperature is 15°C, emergence can be expected in about 7 days if seeds are sown at about 4 cm in a moist seedbed.

Fig. 9.1 *Germination* (0.5).

Leaf emergence (Chapter 3)

Following seedling emergence, there is a delay before the first leaves emerge. During this time the hypocotyl is straightening, leaves in the terminal bud are lengthening and the cotyledons are unfolding. A leaf is said to have emerged when it has started to unfold from the terminal bud and the leaflets have begun to diverge from one another (*Figs* 3.5, 9.4). This is unambiguous compared to, say, referring only to fully expanded leaves which involves subjective decisions (e.g., how does one identify a fully expanded leaf?). Leaves 1 and 2 generally emerge from the terminal bud at much the same time, followed by leaves 3 and 4, then, in sequence 5, 6 and so on; leaves are numbered from the bottom upwards, excluding the cotyledons.

Stem elongation (Chapter 5)

Due to short internodes, young plants usually have a conical shape and leaves must be parted to view the internodes (*Fig.* 9.5). With subsequent development, increases in internode lengths lead to an increasingly elongated plant with the bases of more and more leaves clearly separated by internodes (*Fig.* 9.6) which can therefore be seen without parting the leaves. Eventually the inflorescence bud will be visible without parting the developing leaves (*Fig.* 9.7).

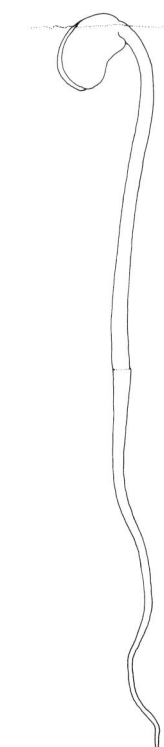

Fig. 9.2 *Emergence* (0.9).

Plants form considerably longer internodes and hypocotyls if grown in low light or in glasshouses. Initially they do not have a conical shape and internodes can be visible without parting leaves. The state of stem elongation can be a comparatively subjective assessment of plant development stage, particularly

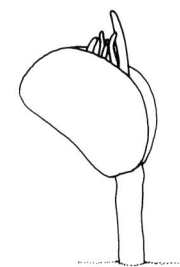

Fig. 9.3 *Leaves protruding beyond cotyledons* (1.0).

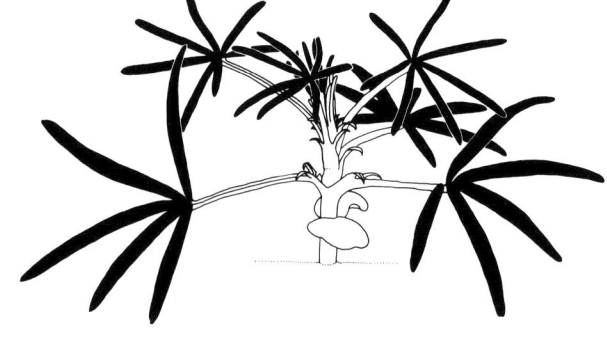

Fig. 9.4 *Seven leaves emerged* (1.7).

Fig. 9.5 *Young, conical-shaped plant* (1.9, 2.1).

Fig. 9.6 *Elongating stem* (1.13, 2.5).

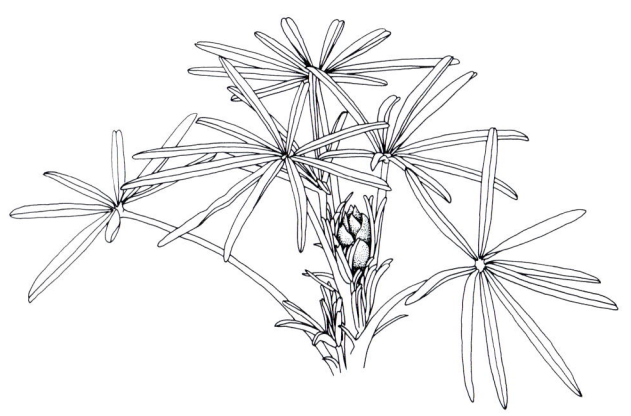

Fig. 9.7 *Inflorescence bud visible (2.7).*

Fig. 9.8 *Peduncle of inflorescence lengthening (2.9).*

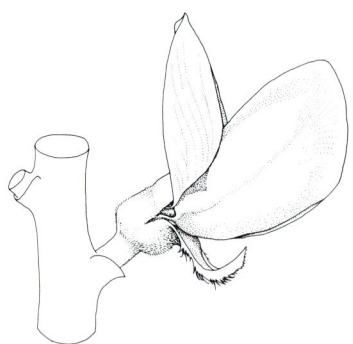

Fig. 9.9 *Open flower (3.4).*

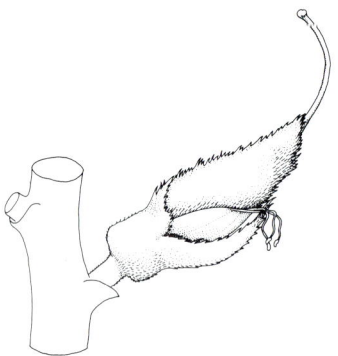

Fig. 9.10 *Pod set (3.9).*

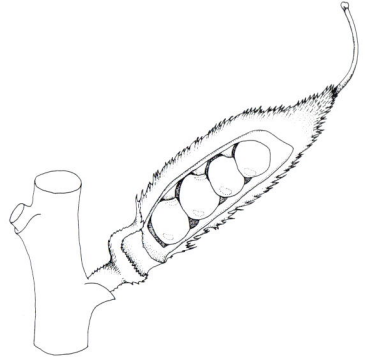

Fig. 9.11 *Young pod without septa, seeds abutting (4.0).*

in view of its sensitivity to the local light environment (hence plant density), so this growth stage criterion should be qualified from another part of the code, such as leaf development or flowering status.

FLOWERING (CHAPTER 6)

Flower development begins with floral initiation (Chapters 3, 5) which is detectable only with microscopic dissection (Chapter 11, *Fig. 5.1*) and occurs when fewer than half the leaves on the main shoot have begun to open, before appreciable stem elongation has occurred (*Figs 1.1, 9.5*). Subsequent leaf opening, and formation and accumulation of flowers on the terminal bud, eventually lead to a clearly visible inflorescence bud (*Figs 9.7, 9.8*). Further development leads to flowering, the flowers opening in sequence up the inflorescence.

The stages of flower opening and subsequent senescence are presented in Table 6.1 in which the flowering category of the scale refers to the stage of development of only the lowest flower. The beginning of flowering for an inflorescence is defined as the attainment of open flower stage of the lowest flower in the inflorescence (raceme) (*Fig. 9.9*). For a crop (or sample of plants from a crop) the beginning of flowering is defined as when half the plants have reached the open flower stage. At this time the flower has just undergone anthesis. The open flower stage is characterised by an erect standard petal (*Fig. 9.9*). Occasionally the standard petal does not become erect, in which case the observer would need to resort to the next stage, when the corolla becomes coloured. Many flowers abscise and some set pods. A pod has set when its tip protrudes through the enclosing petals; this corresponds to a length of about 8–10 mm (*Fig. 9.10*) and few pods abscise after this stage. Occasionally the young pod may remain lodged within the withered petals as the corolla is torn away at its base by the expanding pod.

To extend the description of flowering to a whole inflorescence, the proportion of the inflorescence that has flowered could be assessed. To assess the whole plant, the highest branch order with open flowers or the lowest branch order where flowering has finished might be assessed. Hence such terms as 'flowering on third-order branches' or 'flowering completed on first-order branches', might be used.

POD AND SEED DEVELOPMENT (CHAPTER 7)
Pod and seed development is described for the lowest pod on an inflorescence. Early in pod development, septa, or crosswalls form between the seeds in the pod and the seeds grow to fill the cavity between the septa (seen by a longitudinal section of the pods, *Fig. 9.12*). The septa remain until broken by seed swelling (*Fig. 9.13*) and seeds then touch each other. Pod walls change colour as they mature.

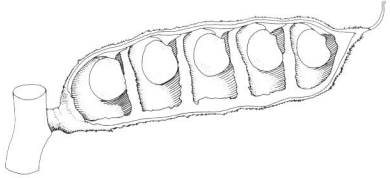

Fig. 9.12 *Septa between seeds, seeds occupying about half the space (4.2).*

The seed initially contains watery, jelly-like endosperm (*Fig. 7.5*) which is released if squeezed. The endosperm is progressively used up as the embryo grows and little if any remains at physiological maturity (*Fig. 7.5*). Colour changes of the seed coat (testa) and the embryo (mainly cotyledons) are indicative of stages in seed development (*Fig. 7.6*). Physiological maturity is the time when seeds reach maximum dry weight and it can be easily recognised by peeling away the seed coat; cotyledons have lost their greenness and are turning yellow to golden orange. At this stage the seeds contain about 62 per cent water (on a dry weight basis); the seeds touch each other; the seed is soft, being easily dented by a finger nail; and may become speckled (*Fig. 7.6*). Later, the seeds shrink rapidly and harden as they lose water. At harvest maturity it is no longer possible to dent the seed with a finger nail or peel away the seed coat and seeds contain about 12 per cent water.

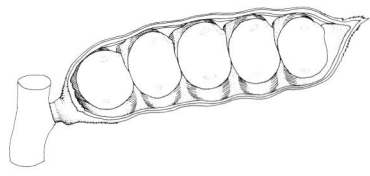

Fig. 9.13 *Septa have split, seeds occupy all the space (4.5).*

To extend the description of pod and seed development to the whole inflorescence, the proportion of pods on an inflorescence with seeds at physiological or harvest maturity could be assessed. Physiological or harvest maturity of the whole plant would be when more than 90 per cent of pods on the highest branch order have reached those stages. In practice, there is only a small gradient in development close to maturity, so assessing the proportion of pods on the plant that contain seeds at physiological or harvest maturity would generally be sufficient. When this figure reaches 90 per cent the plant could be regarded as being at physiological or harvest maturity. A crop may be considered to have reached these stages when 90 per cent of the plants have reached them.

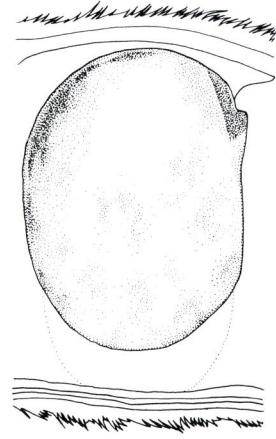 

Fig. 9.14 *Large seed at physiological maturity (5.7).*

Fig. 9.15 *Shrunken, hard seed at harvest maturity (5.9).*

—10—
Predicting Lupin Development

Plant growth and development models can be useful to farmers, agronomists, scientists, and policy makers. Complex models for lupins are not yet feasible, but functions have been developed for predicting emergence, leaf initiation, leaf emergence, vernalisation, leaf numbers, and time to flowering. Functions can be combined for complex investigations, such as for land use suitability.

A MODEL IS A useful aid for the grower making tactical or strategic decisions and for scientists and decision makers exploring complex interactions between plants and the environment. Models have an important role in extrapolating the effects of treatments or responses of genotypes to situations where no direct observations have been made, obviating the need for exhaustive collection of data. On a global scale, lupins are a minor crop and consequently there has been relatively little research on their growth and development physiology. Development of complex models for lupins is, therefore, not yet feasible.

The simple physiological models (functions) described in this chapter are based on few experiments and must be regarded as first attempts which require confirmation from other, independent experiments. The justification for this chapter is that crop modelling in general is a useful tool for crop husbandry and for formulating cropping policies as well as an important discipline for guiding further scientific work. We hope that this chapter will indicate the potential for modelling lupin development and will provide a framework for continued experimental work and model development.

Construction of Crop Models

Models for predicting growth and development have been made for some globally important crops such as wheat and soya beans. These are based on analysis of experiments that relate growth or development to environmental factors. In the case of developmental processes such as flower initiation or leaf emergence, the major environmental factors are generally temperature and daylength, although other factors such as water supply, nutrition and plant density may also affect development.

For example, leaf emergence may be summarised by a function:

$$L = a + bt$$

where L is the number of emerged leaves and t is thermal time. The values for the coefficients a and b are dependent on cultivar and other environmental factors such as daylength and date of sowing. Knowing the coefficients a and b and the value of the variable t, it is possible to calculate the number of emerged leaves.

Functions describing processes such as leaf emergence may be used on their own, but more often they are combined with other similar functions to model the interrelationship between several processes. Thus a function to describe leaf emergence may be part of a calculation of total leaf area of the plant. This, combined with a function to describe photosynthesis in response to the strength of the light, may be part of a model to estimate the growth rate. In such cases the calculations become complex and time consuming, and functions are incorporated in a program which can be run on a computer.

Application of Models

Models may be used for strategic (long-term) or tactical (short-term) purposes or may help to diagnose problems of crop development. An example of the strategic use of models is assessment of the importance of global warming on future farming trends. A model may be used to answer the question "If average temperature increases by 3°C, will it still be profitable to grow wheat (or lupin)?" The answer to such a question may influence policy on future plant breeding programs or research on alternative crops.

Another example, more relevant to lupins, which is an infant crop still expanding into new areas, is "Will lupins succeed in a particular area where they are not grown at present?" Such an exercise has recently been done for Albus Lupin in England and Wales and is described below.

The tactical use of models is usually concerned with the allocation of priorities on the farm. For example, the time for spraying for weed control might be approaching. The most effective time depends on crop stage and the crop may be damaged if the herbicide is applied at the wrong stage. The optimum stage can be identified by plant dissection or number of emerged leaves. The crops under consideration have been sown at different times and with different cultivars. A model that can predict the number of emerged leaves or apex development can assist in deciding the time and order in which the crops should be inspected and sprayed.

Another use of models, combined with crop monitoring or checking, is to identify husbandry problems. For example a Narrow-leafed Lupin crop has been sown, but seedling emergence appears to be delayed. Is anything wrong? If maximum and minimum temperatures are available, thermal time from sowing can be calculated. If the crop has not emerged in about 100°Cd from sowing, or in the case of seed sown into dry soil, from the date of the first adequate rains, then there may be a husbandry or seed problem. Inspection of sowing depth, soil conditions and seed quality may identify the cause.

Lupin models

Seedling emergence
If sown in optimum conditions the time to seedling emergence depends on temperature (*Fig. 10.1*).

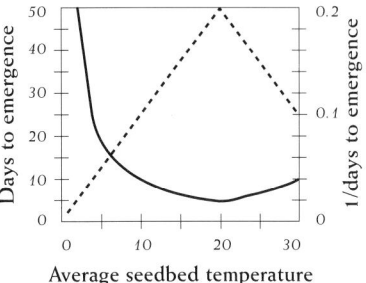

Fig. 10.1 *Relationships between temperature and the number of days to emergence of Narrow-leafed Lupin (———) and the reciprocal of days to emergence (----).*

There is a minimum temperature below which no emergence occurs, as well as an optimum and a maximum temperature (Fig. 10.1). If the reciprocal of days to seedling emergence between the minimum (about 0°C in Narrow-leafed Lupin) and the optimum (about 20°C in Narrow-leafed Lupin) is plotted against temperature, there is an approximately constant response (Fig. 10.1). The thermal time from sowing to emergence is 100°Cd. This can be summarised as:

$$y = 100/t \qquad \text{Eqn 10.1}$$

where y is days to seedling emergence and t is mean temperature. For example, if the mean temperature is 15°C, seedlings will emerge in about 7 days.

Leaf initiation

If the number of leaf primordia is counted at regular intervals and plotted against thermal time, the rate of leaf initiation is approximately constant (Narrow-leafed Lupin, Fig. 10.2a) or may increase slightly with thermal time (Albus Lupin, Fig. 10.2b). In the case of Albus Lupin, seed size (weight) also affects the rate of leaf initation.

Examples of the functions and coefficients are:

For Narrow-leafed Lupin

$$y = 6 + 0.04t \qquad \text{Eqn 10.2}$$

For winter Albus Lupin (300mg seed)

$$y = 4.6 + 0.036t + 0.0000085t^2 \qquad \text{Eqn 10.3}$$

where y is number of leaf primordia and t is thermal time (with base temperature of 3°C for Albus Lupin).

Using these types of functions the thermal time to floral initiation (t) can be calculated if the final number of leaves (y) is known, or if t is known, y can be calculated.

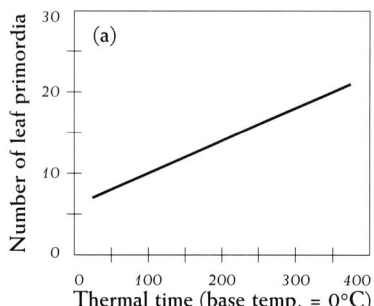

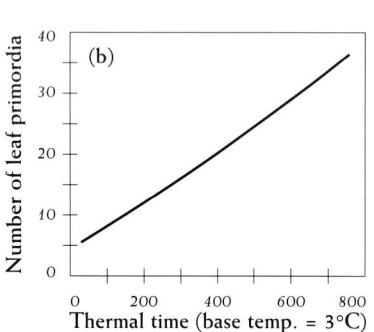

Fig. 10.2 *Appearance of leaf primordia on a thermal time basis for (a) Narrow-leafed Lupin and (b) winter Albus Lupin.*

Vernalisation

Some cultivars of lupin require vernalisation, that is they require a period of low temperature before they will flower.

The critical temperatures and the duration of the required exposure to low temperatures can be quantified.

For Albus Lupin temperatures between 1° and about 14°C are effective (Fig. 10.3). The duration of the period necessary for full vernalisation can be measured in thermal time by summing the daily exposure to vernalising temperature until the plant is fully vernalised (assumed to be when it becomes florally initiated).

If the maximum and minimum daily temperatures fall between the limits for vernalisation then the formula:

$$v = 14 - (max + min)/2 \qquad \text{Eqn 10.4}$$

calculates the daily contribution of vernalising thermal time, where v is the daily vernalising thermal time and max and min are the daily maximum and minimum temperatures, respectively. If the plant is exposed to temperatures outside the vernalisation limits a calculation method to allow for this should be used.

Typical vernalisation requirements for autumn-sown Albus Lupin cultivars range from 300 to 650°Cd.

Leaf emergence
Rates of leaf emergence can be estimated as a function of thermal time as for leaf initiation (Fig. 10.4).

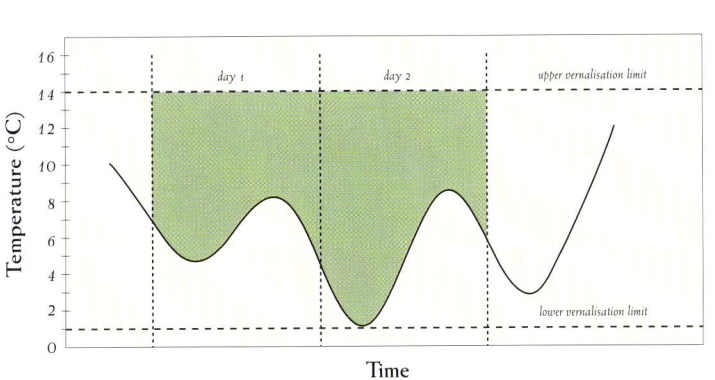

Fig. 10.3 *Diagram to illustrate the temperature limits for vernalisation and the calculation of vernalising thermal time in Albus Lupin. The green area is the cumulative thermal time for days 1 and 2.*

The response of number of emerged leaves to temperature can be summarised as:

$$L = a + bt \qquad \text{Eqn 10.5}$$

where L is the number of emerged leaves and t is thermal time. Typical values of a and b for Narrow-leafed Lupin cultivar Gungurru are −3.7 and 0.03 respectively (base temperature 0°C). If the final number of leaves is known then the thermal time when all leaves have emerged can be calculated.

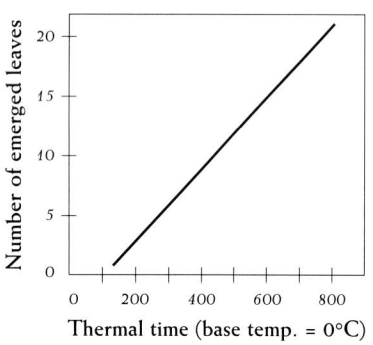

Fig. 10.4 *Appearance of leaves on a thermal time basis (base temperature 0°C) for Narrow-leafed Lupin.*

Time to flowering

Regression models are used to estimate the response of events such as flowering to daylength and temperature. Multiple linear regression of the rate of flowering development (1/days to flowering) against mean temperature and daylength from sowing to flowering gave a good fit to the data:

$$1/d = a + bt + cd \qquad \text{Eqn 10.6}$$

where d is days to flowering, t is mean temperature and d is mean daylength from sowing to flowering.

Typical values for a, b and c for Narrow-leafed Lupin cultivars are shown in Table 10.1

For example, Geraldton, Merredin and Esperance in Western Australia range over five degrees of latitude and have different temperature regimes. Using the model the predicted time of flowering from a common sowing date can be calculated, using long-term temperature data (*Table 10.2, Fig. 6.2*).

AN EXAMPLE OF THE APPLICATION OF A LUPIN MODEL

Land and climate suitability for an Albus Lupin cultivar

Albus Lupin is not grown in England and Wales, but trials of new winter cultivars from France have given good yields and the prospect of commercial cultivation has been considered.

Results from trials have shown that highest yields were obtained from sowings early in September when the plants formed 25–35 leaves on the main shoot. If more than 35 leaves

Table 10.1 *Regression coefficients for Narrow-leafed Lupin cultivars in Equation 10.6*

	PARTIAL REGRESSION COEFFICIENTS		
CULTIVAR	**a**	**b**	**c**
Danja	−0.0308	0.000852	0.00272
Gungurru	−0.0299	0.000636	0.00288

Table 10.2 *Predicted number of days to and date of flowering after sowing the Narrow-leafed Lupin cultivar Gungurru on 29 April.*

	ESPERANCE	GERALDTON	MERREDIN
Days to flowering	95	75	98
Flowering date	2 August	12 July	4 August

were produced the plants were too tall and were prone to lodge (bend down with wind or rain). If too few leaves were produced, yield was reduced and the plants were susceptible to frost damage. The final number of leaves on the main shoot depended on the number of leaf primordia initiated before the plant was fully vernalised (*Fig. 10.5*).

Long term maximum and minimum temperatures were available from meteorological stations distributed around the country. Suitability for growing Albus Lupin was analysed within 55 kilometre squares in a country-wide grid. For each district the time taken to accumulate the 650°Cd of vernalising thermal time to reach floral initiation was claculated for a range of possible sowing dates. The number of leaves initiated during that time was also calculated, and thus the range of sowing dates identified that would lead to plants with 25–35 leaves on the main shoot. For the crop to have a suitable sowing window, the appropriate sowing dates could not conflict with harvest of the preceding crop, thus sowing after the end of August was considered ideal.

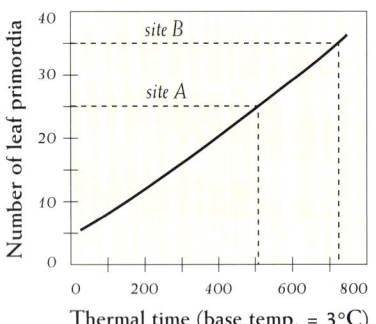

Fig. 10.5 *Model to estimate the final number of main stem leaves on an Albus Lupin using temperature records. At site A, full vernalisation was reached when thermal time for leaf initiation was 510°Cd; the final number was therefore 25. At site B, full vernalisation was reached when thermal time for leaf initiation was 725°Cd; the final leaf number was therefore 35.*

The estimates of number of leaves were superimposed on a soil map. Albus Lupin grows best in soil with a pH of from 4.5 to 7.0. If a square was scored as well suited with respect to number of leaves and the soil pH was between 4.5 and 7.0 then it was considered well suited to lupins. Squares which had more or fewer leaves or where the pH was outside the optimum range received lower scores.

Finally each location was assessed for number of 'machinery working days' in the autumn (which depended on rainfall and soil type) so that the final prediction of suitability for lupin culture was reached taking into account final number of leaves (yield potential), soil pH and number of autumn machinery working days.

From these data a map of England and Wales was constructed showing the predicted suitability for growing a cultivar of Albus Lupin. (For a full explanation of the model consult the paper by Siddons et al. which is listed in the bibliography.)

—11—
Techniques for Examining Lupins

Lupin crops should be regularly inspected (or 'monitored') throughout the season in order to assess the progress of growth and development, and the occurrence of weeds, disease, insects, nutrient deficiencies and adverse environmental conditions (e.g. waterlogging or drought). Proper sampling techniques are necessary to ensure that a representative sample of plants is examined and records of important stages should be kept.

PLANT DEVELOPMENT STAGES should be assessed from a 'typical' plant in the crop, following a planned sampling procedure, such as walking a W path through the crop to sample plants for examination. Some stages such as seedling emergence and flowering can be estimated by inspecting the plant *in situ*. For others, such as floral initiation, plants must be removed for detailed examination. Dissection techniques are necessary to determine such stages as floral initiation.

Long-term or current weather data are necessary to predict or analyse plant development and care should be taken to site the instruments correctly. Written records can be supplemented with photographs. If a crop is monitored regularly, it is helpful to keep records of key stages.

Some events such as seedling emergence occur over several days and the 50 per cent level is usually selected to define the time of an event of this sort. Graphic and other methods can be used to estimate the time of the 50 per cent point.

Regular monitoring of crops is important for: a) timely identification of potential problems such as pests and diseases; b) providing a basis for making more informed management decisions; c) analysing effects of management practices on crop performance; d) allowing growers to share experiences and assess the performance of their crops through comparing performance between paddocks, farms or regions. 'The best manure is the farmer's foot.'

Sampling

Seeds

To ensure that only high quality seed is sown, seed testing is important, especially for germinability, seed weight, virus infection and seedling vigour. Thorough sampling is essential. Many small samples (about 100) should be taken from each 25 tonnes of seed and amalgamated to achieve a sample of approximately 20 kg. The small samples could be removed using a scoop or a cup placed in the seed stream at regular intervals while seed is being transferred to a bin, truck or silo. An alternative but less preferable method is to remove seeds using a sampling spear in the bin, truck or silo. In this case the full depth of seed should be sampled at about 30 evenly spaced points for 25 tonnes of seed.

The amalgamated seed samples should be reduced in volume to finish up with about 2 kg. This is done by mixing the amalgamated sample thoroughly, dividing it into halves, taking one half and repeating the process until about 2 kg remain. This, in turn, will be subsampled for the required seed analyses.

Plants

Observations on developmental stages such as emergence, flowering and maturity can be made *in situ*. Stages such as floral initiation require destructive sampling and it is impractical to assess large numbers of plants. Because of variation among plants and throughout a paddock, any estimate of an 'average' or 'typical' stage of plant development must be based on a representative sample from the crop or experimental plot. Observations should therefore be made throughout the paddock. A convenient way to locate the observations is to walk a 'W' route across the paddock, taking measurements at points along each leg of the W. Areas of poor growth for which the reason is known should be avoided, for example a compacted area near a gateway. Usually 10 observation points are sufficient. In the case of designed field experiments, observations should be made at random within each plot, leaving a 'border' or 'guard strip' around observation areas.

To aid in diagnosing unknown problems, it is sometimes worthwhile to sample plants from areas of poor growth and compare them with plants sampled from nearby areas of good growth.

In situ observations. For non-destructive observations, at each observation point, an area of ground (quadrats) should be marked out for recording plants that have reached the stage of interest. A convenient method is to use a quadrat of one square metre and count the number or proportion of plants which have reached the stage of interest. For lupins sown in rows, the dimensions of the quadrat should be compatible with the row spacing, for example if rows are 18 cm apart, the quadrat could cover 4 rows (hence be 72 cm wide) and by making it 139 cm long the area enclosed would be one square metre (*Fig. 11.1*). Alternatively, plants could be assessed within 5 or 6 separate metre lengths of row (each metre of row equates to 0.18 square metres if rows are 18 cm apart) at each observation point.

To identify accurately when half the plants have reached the stage of interest, regular observations should be made and the quadrats should be fixed so that the observations are made on the same areas. Shoots may become intermingled, in which case observations are made only on those plants whose main shoot originates within the quadrat. Recording and calculating seedling emergence is explained below, in 'Recording' and is similar for other developmental stages such as flowering, or physiological or harvest maturity.

Fig. 11.1 *Seedling emergence can be recorded using one square metre quadrats placed in several positions in the crop.*

Fig. 11.2 *Soil can be carefully removed from around roots to reveal where problems with root growth are occurring. The tip of this tap root is thickened and distorted indicating it has reached a compact layer.*

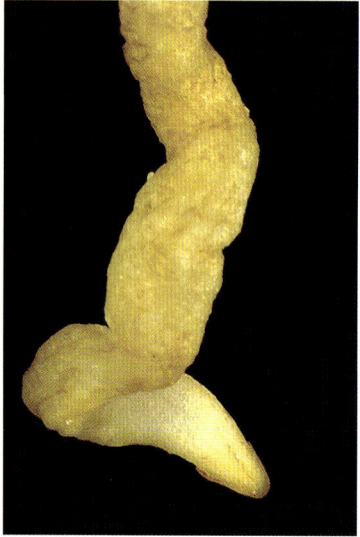

Fig. 11.3 *Thickened and distorted root tip, characteristic of a root trying to grow through a compact layer.*

An alternative but less desirable method for monitoring development progress (except for emergence) is to observe only a few (say, 5) tagged plants rather than all the plants in a quadrat. This method may also be used when it is not practical to observe so many plants, such as if making progressive leaf counts

Destructive observations. For assessing depth of sowing or nodulation, ten seedlings should be dug up at each observation point within a field or paddock. The seedlings are most conveniently dug up from different rows using a trowel, ensuring that about 10 cm of root remains with the plant for nodulation assessment. Depth of sowing is indicated by the length of white hypocotyl (*Fig. 2.4*) and is best assessed shortly after seedlings have ceased to emerge.

If a detailed examination of the plant is to be made, it is often easier to do this at a table or bench rather than in the field. Plants should be dug up carefully at each observation point, put in a polythene bag to prevent drying out and kept as cool as possible. If examination is delayed, then storage in the salad region of a refrigerator with wet tissue in each bag will keep plants fresh and turgid for one to two days, during which time there will be no appreciable change in development stage. If development is to be assessed on only a subsample of plants removed from the field, it is best to arrange plants in order of size and use plants from the median class.

ROOT INSPECTION

It is sometimes helpful to examine the root system *in situ*, such as when trying to diagnose reasons for poor shoot growth. The simplest method is to dig a small trench parallel to and about 10 cm from a crop row, before the soil starts to dry out. Gentle removal of soil from the face of the trench closest to the crop row will expose roots (*Fig. 11.2*). Characteristics of the roots to observe include presence or absence of disease, root death, localised thickening, distortion (*Fig. 11.3*) or accumulation of roots (which might indicate the top of a compacted layer, 'plough pan'), colour of young roots and depth of penetration. Additional observations of depth of the soil profile (down to hard clay or rock) and depth to a raised water table are often also useful.

Occasionally a deep trench will yield instructive information. This will necessitate using a back hoe to create a pit, up to 3 m deep. By standing in the pit, soil can be gently removed from a face to expose roots. Care is needed since the sides of the pit could collapse.

PLANT DISSECTION

Some important development stages cannot be seen with the naked eye because the apex is too small and must be dissected to remove the enfolding leaves. A microscope and a few instruments are necessary for plant dissection (*Fig. 11.4*). These are:

- a stereoscopic dissecting microscope with a magnification range of about 5–20 times

- a fine needle, preferably in a chuck handle so that it can be replaced. Fine sewing needles are best

- a pair of fine forceps

- a scalpel with fine, replaceable blade.

- some modelling clay (for example, Plasticine or Bluetak), useful for holding small fiddly bits of a plant while dissecting.

SHOOT APEX

Remove the leaves with fingers or forceps. If the number of leaves on a stem is being counted and some leaves have already fallen, find and include leaf scars in the count (remembering that the cotyledons also leave scars when they fall off). When it is difficult to see the smaller leaves, the shoot is transferred to the microscope and the remaining leaves are removed under low magnification e.g. x5 or x10. Small lupin leaves, particularly those of Albus Lupin, are woolly and felted together and are easier to remove with fine forceps than with a needle. Finally the exposed apex is examined under high magnification e.g. x50. The number of leaves and whether the apex is vegetative or floral (Chapters 3 and 5) may be recorded, depending on the purpose of the examination. If the date of floral initiation is to be determined, frequent

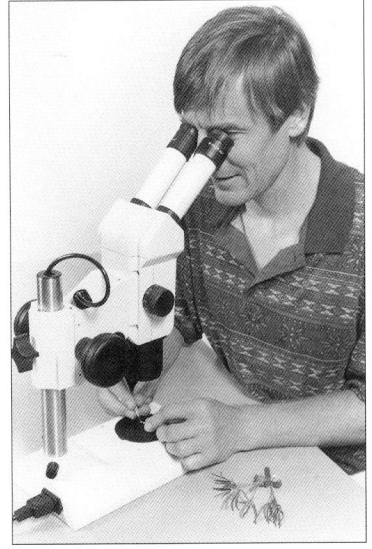

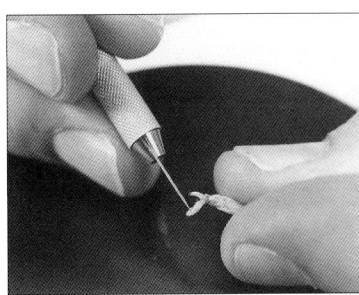

Fig. 11.4 *Plant dissection requires a stereomicroscope (a) and instruments, such as a fine scalpel or needle for carefully removing developing leaves (b).*

dissections are desirable. If the final number of leaves, typical of the cultivar under investigation, is known, then occasional leaf counts by dissection will give clues as to when intensive sampling is necessary. Otherwise, look for changes in the shape of the shoot apex (Chapter 5). Dissection should continue after the first floral primordia are seen, as the early stages of flower development are sometimes difficult to distinguish.

SEED

A mature dry seed is very difficult to dissect and examine, e.g. for damage to, or faulty development of, the embryo. If the seed is soaked in water near to freezing point for a day or so the seed will take up water but will not grow appreciably. The seed coat can then be cut and the embryo removed without damage.

FLOWER

Immature flowers can be examined while attached to the rachis. First the bract is removed, then the petals and sepals are picked off with forceps to examine the stage of development of the stamens and ovary (see *Figs 5.5–5.8* for details of flower structure). As the flower nears anthesis, care is needed to prevent anthesis (rupture of the anthers) being induced by dissection, thus giving a false impression of the stage of flower development.

POD AND SEED

The condition of the pod and seed can give useful information about progress to maturity (for example *Fig. 7.4*). If the pod is cut longitudinally, in the vertical plane (see *Fig. 7.3*) the appearance of the septa (cross-walls) and the relative volumes of the seed and the pod cavity can be assessed (development stage scale, Chapter 9). If the immature seed is cut around the circumference and the seed coat lifted off, the condition of the embryo and presence of endosperm (if any remains) can be observed (for example *Fig. 7.5*).

Colour changes in the pod and seed are useful indicators of progress to maturity and can be matched against the colour diagrams (Chapter 7).

WEATHER RECORDS

Rainfall records are important. If the crop is sown dry, then germination is not possible until the first adequate rains, so this becomes the effective sowing date. In Mediterranean-type environments, growing season rainfall is the main determinant of crop yield, so rainfall records can help identify why crop yield potentials may not be reached and can also explain some of the season to season variation in yield.

Temperature is an important determinant of the time to flowering and other stages of lupin development. Long term temperature records can therefore give predictions about forthcoming stages of development (Chapter 10). To relate crop development to temperature or to identify problems associated with high or low temperatures, current temperature records are necessary. If there is no weather station nearby from which temperatures can be obtained, maximum and minimum temperatures can be recorded but care should be taken to site the instruments properly. Ideally, the thermometers should be in a well ventilated white box about 1.5 m above the ground (Stevenson Screen), away from buildings and trees and near but not in the crop.

PHOTOGRAPHY

Photography is a good way to keep records of crop development or to convey problems to others for identification. A x2 macro lens, available for most single lens reflex cameras, is adequate for taking informative pictures of individual flowers and other small plant parts. If accurate colour rendition is desired, light quality is important. Photographs should be taken in full sun or with light cloud. Photography should be avoided within two hours after dawn or before sunset, as at those times the light has a red cast. Careful focusing and exposure at a wide f-stop (about f 2.8) will blur the background and accentuate the features of the subject. Plants can also be dug up and photographed in isolation. A pale blue or fawn coloured stiff card makes a good background for green plants, grey is suitable for ripe plants.

Recording

The following cardinal points provide a basic record of plant development (bracketed stage requires some extra preparation and equipment):

- sowing date
- seedling emergence
- (floral initiation)
- inflorescence bud
- anthesis
- pod set
- physiological maturity
- harvest

Stages such as seedling emergence and flowering are usually defined by the 50 per cent level. For example, to establish 50 per cent seedling emergence, frequent counts are made of seedlings emerged in a quadrat. The number of seedlings in each of a number of fixed quadrats placed at representative positions in a crop (on the segments of a W) is counted from the emergence of the first seedling until all the seedlings have emerged (when the number remains static over two or more counts). Mean values for each visit are calculated and plotted versus time or thermal time (*Table 11.1, Figs 11.5 and 11.6*). The 50 per cent value can then be read off the graph. (There are a number of sophisticated methods available for analysis of data of this type). Similar methods may be employed to score other events such as flowering or physiological maturity. For recording the end of stages, such as flowering the 90 per cent value is generally used, for example when flowering has finished in 90 per cent of plants.

Chapter 11: Techniques for Examining Lupins

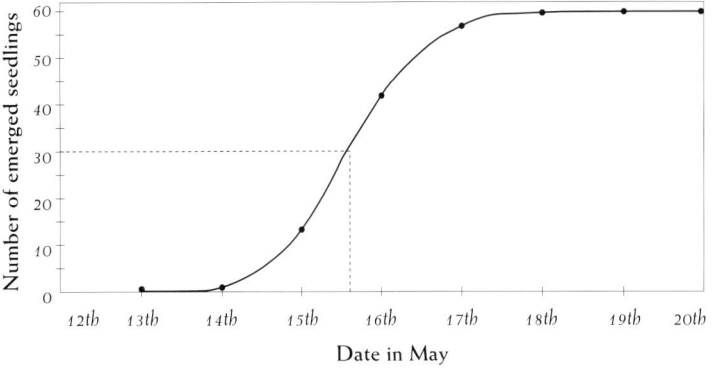

Fig. 11.5 Mean number of emerged seedlings from Table 1, plotted against date. Observed values (●). The dashed lines show 50% emergence, which occurred on 16 May (X-axis marks are at midday).

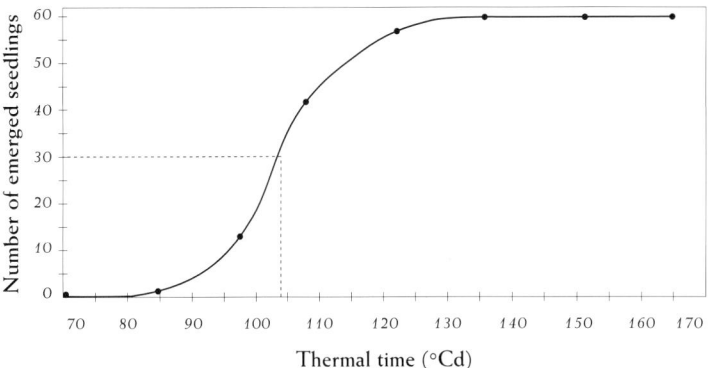

Fig. 11.6 Mean number of emerged seedlings from Table 11.1, plotted on a thermal time basis. Observed values (●). The dashed lines show 50% emergence, which occurred a little after 100°Cd. This is close to the predicted value, indicating that the seeds were healthy and sown in good consitions (correct depth with adequate water).

Table 11.1 Daily counts of number of emerged seedlings in two, half-square metre quadrats for ten observation points from 12 to 20 May. Maximum and minumum temperatures and thermal time are shown. The mean maximum number of seedlings was 60, thus at 50% emergence there were 30 seedlings. Seeds were sown on 7 May.

Date	Temperature (°C) Max.	Min.	Thermal Time (°Cd)	Number of Emerged Seedlings at Each Observation Point						
				1	2	3	...	9	10	Mean
12 May	16.6	5.3	53				...			
13 May	19.7	14.6	70	0	0	0	...	0	0	0
14 May	19.7	9.3	85	1	0	1	...	1	2	1
15 May	18.4	7.3	97	15	8	12	...	16	11	12.4
16 May	13.2	7.7	108	45	31	44	...	48	41	41.8
17 May	18.7	10.3	122	54	52	62	...	60	56	56.8
18 May	20	7.2	136	56	57	64	...	61	60	59.6
19 May	19.4	11	151	56	58	64	...	61	61	60
20 May	19.8	7.8	165	56	58	64	...	61	61	60

List of Abbreviations

cm	centimetre	pH	a measure of acidity; pure water has a pH of 7 (neutral), acidic solutions have lower values and alkaline solutions higher values.
kg	kilogram		
km	kilometre		
m	metre	sq.	square
mg	milligram	°	degree (of the compass)
mm	millimetre	°C	degree Celsius
MPa	megapascal, a unit of pressure	°Cd	thermal time; see Glossary
		%	per cent

Glossary

Note: a number of terms are defined here in the sense used in this book.

abortion: arrested development (of pod or seed) leading to sterile, shrunken organ.

abscission layer: one or more layers of thin-walled cells which may form across the stalk of a plant part, such as the pedicel of a flower; breakdown of the bonding material between these cells leads to shedding of the plant part.

adventitious root: a root originating from an organ other than another root, for example the hypocotyl.

anther: see **stamen**.

anthesis: the time of anther dehiscence, also used to indicate the time of full expansion of a flower.

apical dome: a convex grouping of dividing cells at the apex of a shoot, from which the precursors of tissues arise.

axil: the angle between a stem and a leaf.

axillary bud: a bud that forms in the axil between the stem and a leaf; *in lupins*, consisting of a meristematic apex, with or without young leaves and/or flowers which have little vertical separation.

basal/lower branch: a branch arising in the axil of the cotyledons or lowest 4–6 leaves on the main shoot.

bract: a leaf-like structure that covers a flower in the early stages of its development; *in lupins*, attached at the base of the pedicel

bud: a group of meristematic cells for the initiation of vegetative or reproductive structures; may also include young leaves or flowers with little vertical separation.

calyx: the green outer structures of a flower; made up of the sepals.

canopy: the cover provided by a plant's foliage.

chalaza: the region in the ovule where the nucellus and the integuments connect with the funicle.

corolla: the parts of the flower between the calyx and the sexual organs; made up of the petals.

Glossary

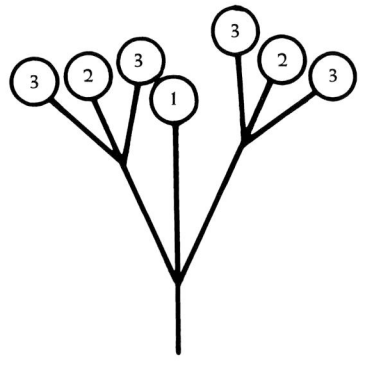

cyme

cotyledons: *in lupins,* a pair of thick, leaf-like food storage organs that account for most of the mass of mature seeds; after germination they are carried toward the soil surface by elongation of the hypocotyl and become green after emerging from the soil.

cyme: an inflorescence in which the apex of the main stem produces a single flower; further flowers develop at the end of lateral branches below it, followed by more flowers at the end of lateral branches below them. See diagram, which also shows the order of flowering.

dome: see **apical dome**.

embryo: *in lupins,* the young, partially developed, plant found in the mature seed, consisting of an axis—the hypocotyl-root axis—bearing the radicle, hypocotyl, cotyledons and shoot apex.

endosperm: nutritive tissue within the embryo sac; *in lupins,* consumed during embryo development.

epicotyl: the shoot part of the embryo or a seedling above the cotyledons consisting of an axis, young leaves and leaf primordia.

Faboideae: the genus *Lupinus* is a member of the family *Fabaceae* (or *Leguminosae*). The *Fabaceae* are divided into three subfamilies—*Caesalpinioideae; Mimosoideae* and *Faboideae* (or *Papilionoideae*). Like all grain legumes, lupins are a member of the *Faboideae,* which is divided into a number of tribes. *Lupinus* is a part of a distinct group (subtribe Lupininae) within the tribe *Genisteae*. The genus contains about 200 species with two principal areas of distribution; the length of the Americas including the Andes and the Rocky Mountains, and the Mediterranean basin. The species considered in this book—*L. albus, L. angustifolius* and *L. luteus*—are native to the Mediterranean region: *L. albus* at the eastern end; *L.luteus* at the western end; and *L.angustifolius* all around. Now they are all, especially *L. albus,* widely spread and cultivated in various parts of the world.

filament: *see* **stamen**.

first-order branch: a branch originating in a leaf or cotyledon axil on the main shoot.

floral apex: a group of meristematic cells at the apex of a shoot that, by cell division, initiate flowers.

floral initiation: the stage when meristematic cells at the shoot apex switch from initiating vegetative structures to initiating flowers.

floral phase: the phase following the vegetative phase; from floral initiation until anthesis.

flowering: *on a shoot,* the phase following the opening of the first flower while further flowers continue to open.

flower primordium: *see* **primordium**.

funicle: a seed stalk which attaches the ovule (seed) to the placenta in the pod.

genotype: organism(s) with a discrete genetic constitution, in contrast with *phenotype*—the physical characteristics manifested by an organism.

germination: protrusion of the radicle, or primary root through the seed coat; *in this book*, germination is arbitrarily defined as having taken place when the radicle projects more than 5 mm from the seed.

harvest maturity: the stage of a plant's life cycle when mechanical harvesting of seed is feasible.

hilum: a scar on a seed resulting from its abscission from the funicle.

hull/seed coat: outer covering of the seed; *see also* **testa**.

hypocotyl: axial part of the embryo or seedling between the cotyledons and the radicle.

inflorescence: the part of a shoot that bears flowers; *in lupins* a **raceme**, *q.v.*

internodes: the region of a stem between successive nodes.

introrse dehiscence: *of anthers*, split on an inner suture, casting the pollen into the centre of the ring of anthers.

keel petals: see **petals**.

leaf emergence: separation of a leaf from the terminal bud, coinciding with separation of leaflets from each other.

leaf primordium: a localised region of cell division and extension on the flank of the shoot apex from which a leaf originates.

leghaemoglobin: protein involved with transport of oxygen within legume nodules; responsible for the pink colour of nodules.

legume: a dry fruit (pod) containing one to several seeds, derived from a single ovary of one carpel and splitting along one or both margins (sutures).

lodge: to bend down with wind or rain.

main shoot / stem: the first axis of the plant to develop, together with its appendages (leaves, branches, flowers).

meristematic dome: the topmost part of a shoot, containing dividing cells which initiate new organs.

monadelphous: having the filaments of the stamens joined together from the base to form a tube. Some genera in the *Fabaceae* have a diadelphous arrangement in which the posterior stamen is free.

mycorrhizal fungi: fungi having a symbiotic association with a plant through hyphae which penetrate the root.

Glossary

node: the point on a stem at which a leaf arises.

nodule: an outgrowth of tissue from a root, having cells containing nitrogen-fixing bacteria (bacteroids) and leghaemoglobin.

ovary: a structure below the style containing the ovules; develops into the fruit.

ovule: an egg cell within the ovary which, after fertilisation by a male reproductive cell, differentiates into a seed.

palmate: a leaf shape with radiating lobes (leaflets).

pedicel: the stalk of an individual flower, *in lupins*, subtended by a bract.

peduncle: the stem of an inflorescence; the portion between the highest leaf and lowest flower.

petals: the showy parts of a flower within the calyx; *in lupins*, comprising the keel petals (lowest two petals), wing petals (two lateral petals) and standard petal or vexillum (uppermost, distinct and largest petal).

petiole: the stalk of a leaf.

photoperiodism: promotion of flowering by daylength (photoperiod); long-day plants such as lupin flower sooner as the days get longer, while flowering of short day plants such as soya bean is hastened by shorter days.

phyllochron: the time or thermal time from the emergence of one leaf to the next.

physiological maturity: the stage of seed development when maximum dry mass has been achieved; thereafter fresh mass declines as water is lost.

plastochron: the time or thermal time from the initiation of one primordium to the next.

pod set: *in this book*, the time when a pod is clearly visible, 8–10mm long, protruding beyond the withered petals if they are still present or if the corolla has been torn free at the base.

pollen tube: a tubular extension formed as a pollen grain germinates on the stigma; the tube penetrates the stigma and grows down the style, carrying the male gametes to the ovules in the ovary.

primordium: an organ, cell or organised series of cells in a very early stage of differentiation e.g. leaf primordium, flower primordium, bract primordium.

proteoid or cluster roots: a dense proliferation of short roots that are densely covered in root hairs.

pulvinus: a structure at the base of a leaflet that has a role in the movements of the leaflet.

Glossary

pure line: genotype produced by continuous self pollination.

quadrat: a marked ground area in which sampling or observation of plants is undertaken.

raceme: an inflorescence with a simple axis (rachis) bearing flowers on stalks (pedicels) that are subtended by bracts. See diagram, which also shows the order of flowering.

rachis: the main axis of an inflorescence from the lowest flower to the tip and on which the flowers are borne.

radicle: the embryonic root in the seed which protrudes during germination and becomes the tap root; within the seed the radicle is encased in a pocket in the seed coat, the **radicle pocket**.

rhizobia or **bradyrhizobia**: bacteria that infect plant roots and form a symbiotic relationship with the host plant, during which they make nitrogen from the air available to the plant.

rhizosphere sheath: a zone around young roots strongly influenced by root excretions; seen as the sheath of adhering soil particles after careful removal of roots from soil.

root cap: a cap of loosely organised cells covering the tip of a root and protecting the apex as it is forced through the soil.

root hair: a filamentous projection from a single root epidermal cell.

second-order branch: a branch originating in the axil of a leaf of a first-order branch.

seedling axis: *see* **embryo**.

seedling emergence: protrusion of any part of a seedling through the soil surface.

sepals: the leaf-like unit of the calyx; outside the petals.

septum: a partition, e.g. between seeds in a pod.

shoot apex: the top-most part of a shoot, containing the apical meristem (dome) and associated leaf or flower primordia.

silly seedling syndrome: the condition in a seedling when the root appears to grow upward rather than downward.

stamen: male organ of a flower; consisting of pollen sacs (anthers) carried on the end of a stalk called the filament; *in lupins* they are joined at their base to form a tube surrounding the ovary.

standard petal: *see* **petals**.

stigma: *see* **style**.

stipule: small, lateral part of a leaf originating at the leaf base, one on each side.

style: stalk-like projection from the ovary, ending in the receptive female surface called the stigma.

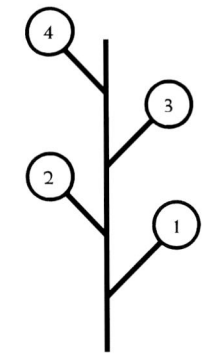

raceme

Glossary

symbiotic nitrogen-fixing bacteria: fixation of atmospheric nitrogen in root nodules; the host plant provides the bacteroids with carbohydrates used to reduce nitrogen to ammonium.

tap root: the prominent main root of dicotyledons, from which the first-order lateral roots develop.

terminal bud: a bud at the tip of a shoot, consisting of the shoot apex, with or without young leaves and/or flowers which have little vertical separation.

testa: the seed coat; differentiated from the integuments of the ovule.

thermal time: the summation of temperatures to quantify or predict the duration of developmental phases. (Equivalent terms are *degree days, heat units, heat sums, thermal units* and *growing degree days*).

The most useful definition of thermal time is $Tt = \sum_{i=1}^{n}(T_a - T_b)$ where Ta is the daily mean temperature, Tb is the base temperature below which development stops, and n is the number of days over which temperature is summed. Ta is usually calculated as the average of daily maximum and minimum temperatures. This method of calculation is appropriate for predicting plant development if the following conditions are met:

1. response of development to temperature is linear over the range experienced
2. temperatures do not fall below Tb
3. temperatures do not exceed an upper threshold (optimum temperature
4. the developing organ (shoot apex, leaf meristem, etc.) is at the same temperature as Ta.

Other more complicated methods of calculating thermal time are available if these conditions are not met.

upper/apical branches: branches arising in the axils of the uppermost 4–6 leaves of a shoot or branch.

valve: a segment of a pod after dehiscence along both margins; in *lupins* the pod wall comprises 2 valves.

vegetative phase: the phase following germination and continuing until initiation of flowers begins at the shoot apex.

vernalisation: promotion of flowering by low temperatures.

vexillum: the uppermost, largest petal of a lupin flower, also termed the **standard**.

water use efficiency: the efficiency with which a plant makes matter (e.g. dry mass or seeds) per unit of water used.

wing petals: See petals.

BIBLIOGRAPHY

Ball, E. (1949), The shoot apex and normal plant of *Lupinus albus* L., bases for experimental morphology, *American Journal of Botany* **36**, 440–459.

Boundy, K.A., Reeves, T.G. & Brooke, H.D. (1982), Growth and yield studies of *Lupinus angustifolius* and *L. albus* in Victoria, *Australian Journal of Experimental Agriculture and Animal Husbandry* **22**, 76–82.

Clapham, W.M. & Barnes, S.L. (1990), Development and maturation of white lupine seed, *Agronomy Journal* **82**, 707–710.

Clements, J.C., White, P.F. & Buirchell, B.J. (1993), The root morphology of *Lupinus angustifolius* in relation to other *Lupinus* species, *Australian Journal of Agricultural Research* **44**, 1367–1375.

Davies, S. & Williams, W. (1983), Rates of pod and seed development in *Lupinus albus*, *L. mutabilis*, *Vicia faba*, *Pisum sativum* and *Lathyrus latifolius*, in: R. Thompson & R. Casey (eds), *Perspectives for Peas and Lupins as Protein Crops* pp. 8–9 (Martin Nijhoff, The Hague).

Davies, S. & Williams, W. (1985), The rate of morphogenesis of embryos and seeds of four species of grain legumes, *Annals of Botany* **56**, 429–435.

Delane, R.J., Hamblin, J. & Gladstones, J. (1986), Reduced-branching Lupins, *Western Australian Journal of Agriculture* **27**, 47–48.

Dracup, M. (1994), Physiological aspects to lupin improvement for water-limited mediterranean environments, in: *Proceedings of the First Australian Lupin Technical Symposium, Perth*, pp. 140–151 (Western Australian Department of Agriculture, Perth).

Dracup, M., Davies, M. & Tapscott, H. (1993), Temperature and water requirements for germination and emergence of lupin, *Australian Journal of Experimental Agriculture* **33**, 759–766.

Dracup, M. & Kirby, E.J.M. (1993), Patterns of growth and development of leaves and internodes of narrow-leafed lupin, *Field Crops Research* **34**, 209–225.

Dracup, M. & Kirby, E.J.M. (1995), Architectural stability in restricted branching narrow-leafed lupin, in: *Proceedings of the Second European Conference on Grain Legumes, Copenhagen*, pp. 46–47 (European Association for Grain Legume Research).

Dracup, M. & Kirby, E.J.M. (1996), Pod and seed growth of narrow-leafed lupin in a water-limited mediterranean-type environment, *Field Crops Research* (in press).

Duthion, C. & Ney, B. (1991), Chronology of reproductive development of four types of white lupin (*Lupinus albus* L.). Period of seed number formation on the different reproductive structures, *in: Proceedings of the 6th International Lupin conference, Temuco-Pucon, Chile*, pp. 255–261, Asociación Chilena del Lupino, Temuco, Chile.

Duthion, C., Ney, B. & Munier-Jolain, N.M. (1994), Development and growth of white lupin: Implications for crop management, *Agronomy Journal* **86**, 1039–1045.

Ehleringer, J.R. & Forseth, I.N. (1989), Diurnal leaf movements and productivity in canopies, *in:* G. Russell, B. Marshall & P.G. Jarvis (eds), *Plant Canopies: their Growth, Form and Function*, pp. 129–141 (Cambridge University Press, Cambridge).

Ellis, R.H., Hong, T.D. & Roberts, E.H. (1987), The development of desiccation-tolerance and maximum seed quality during seed maturation in six grain legumes, *Annals of Botany* **59**, 23–29.

Farrington, P. (1979), A study of the interactions of vegetative growth and fruit setting in *Lupinus angustifolius* cv Unicrop, Ph.D thesis, University of Western Australia.

Farrington, P. & Greenwood, E.A.N. (1975), Description and specification of the branching structure of lupins, *in:* Greenwood, E.A.N., Farrington, P. & Beresford, J.D., Characteristics of the canopy, root system and grain yield of a crop of *Lupinus angustifolius* cv Unicrop, *Australian Journal of Agricultural Research* **26**, 497–510, Appendix 1.

Gladstones, J. (1994), An historical review of lupins in Australia, *in: Proceedings of the First Australian Lupin Technical Symposium, Perth*, pp. 1–38 (Western Australian Department of Agriculture, Perth).

Gladstones, J.S. & Hill, G.D. (1967), Selection for economic characters in *Lupinus angustifolius* and *L. digitatus*. 1. Non-shattering pods, *Australian Journal of Experimental Agriculture and Animal Husbandry* **7**, 360–366.

Greenwood, E.A.N., Farrington, P. & Beresford, J.D. (1975), Characteristics of the canopy, root system and grain yield of a crop of *Lupinus angustifolius* cv. Unicrop, *Australian Journal of Agricultural Research* **26**, 497–510.

Hamblin, A.P. & Hamblin, J. (1985), Root characteristics of some temperate legume species and varieties on deep, free-draining entisols, *Australian Journal of Agricultural Research* **36**, 63–72.

Hamblin, A. & Tennant, D. (1987), Root length density and water uptake in cereals and grain legumes: How well are they correlated?, *Australian Journal of Agricultural Research* **38**, 513–527.

Herbert, S.J. (1979), Density studies on lupins. 1. Flower development, *Annals of Botany* **43**, 55–63.

Herbert, S.J. (1979), Density studies on lupins. II. Components of seed yield, *Annals of Botany* **43**, 65–73.

Huyghe, C. (1991), White lupin architecture—genetic variability—agronomic consequences, in: *Proceedings of the 6th International Lupin Conference, Temuco-Pucon, Chile*, pp. 241–254 (Asociación Chilena del Lupino, Temuco, Chile).

Huyghe, C. (1991), Winter growth of autumn-sown white lupin (*Lupinus albus* L.): main apex growth model, *Annals of Botany* **67**, 429–434.

Huyghe, C., Harzic, N., Julier, B. & Papineau, J. (1994), Comparison of determinate and indeterminate autumn-sown white lupins under the western European climate, in: *Proceedings of the First Australian Lupin Technical Symposium, Perth*, pp. 123–128 (Western Australian Department of Agriculture, Perth).

Huyghe, C. & Papineau, J. (1990), Winter development of autumn-sown white lupin: agronomic and breeding consequences, *Agronomie* **10**, 709–716.

Julier, B. & Huyghe, C. (1993), Description and model of the architecture of four genotypes of determinate autumn-sown white lupin (*Lupinus albus* L.) as influenced by location, sowing date and density, *Annals of Botany* **72**, 493–501.

Ma, Q., Longnecker, N., & Dracup, M., Nitrogen deficiency affects leaf development and delays flowering in narrow-leafed lupin, *Annals of Botany* (submitted).

Milford, G.F.J., Day, J.M., Huyghe, C. & Julier, B. (1993), Floral determinacy in autumn-sown white lupins (*Lupinus albus*); the development of varieties for cooler European climates, *Aspects of Applied Biology* **34**, 89–97.

Nelson, P. & Delane, R. (1990), *Producing lupins in Western Australia*, Bulletin 4179 (Western Australian Department of Agriculture, Perth).

O'Neill, T.B. (1961), Primary vascular organisation of lupinus shoots, *Botanical Gazette* **123**, 1–9.

Pate, J.S. & Farrington, P. (1981), Fruit set in *Lupinus angustifolius* cv. Unicrop II. Assimilate flow during flowering and early fruiting, *Australian Journal of Plant Physiology* **8**, 307–318.

Pate, J.S., Williams, W. & Farrington, P. (1985), Lupin (*Lupinus* spp), in: R.J. Summerfield & E.H. Roberts (eds), *Grain Legume Crops*, pp. 699–746 (William Collins & Sons, London).

Perry, M.W. (1975), Field environment studies on lupins. II. The effects of time of planting on dry matter partition and yield components of *Lupinus angustifolius* L., *Australian Journal of Agricultural Research* **26**, 809–818.

Perry, M.W., Delane, R.J., Tennant, D. & Hamblin, A.P. (1986), The growth and grain yield of lupins in relation to the soil water balance, in: *Proceedings of the 4th International Lupin Conference, Geraldton, Western Australia*, pp. 112–118 (Western Australian Department of Agriculture, Perth).

Perry, M.W. & Poole, M.L. (1975), Field environment studies on lupins. I. Developmental patterns in *Lupinus angustifolius* L., the effects of cultivar, site and planting time, *Australian Journal of Agricultural Research* **26**, 81–91.

Peterson, C.M., Mosjidis, C.O., Dute, R.R. & Westgate, M.E. (1992), A flower and pod staging system for soybean, *Annals of Botany* **69**, 59–67.

Pigeaire, A., Delane, R., Seymour, M. & Atkins, C.A. (1992), Predominance of flowers and newly formed pods in reproductive abscission of *Lupinus angustifolius* L., *Australian Journal of Agricultural Research* **43**, 1117–1129.

Pigeaire, A., Seymour, M., Delane, R. & Atkins, C. (1992), Partitioning of dry matter into primary branches and pod initiation on the main inflorescence of *Lupinus angustifolius*, *Australian Journal of Agricultural Research* **43**, 685–696.

Polhill, R.M. (1976), Genisteae (Adans.) Benth. and Related Tribes (Leguminosae), *in*: V.H. Heywood (ed), *Botanical Systematics*, pp. 143–333 (Academic Press, London).

Polhill, R.M. & Raven, P.H. (1981), *Advances in Legume Systematics*, Ministry of Agriculture, Fisheries and Food, U.K.

Reader, M., Dracup, M. & Kirby, E.J.M. (1995), Time to flowering in narrow-leafed lupin, *Australian Journal of Agricultural Research* **46**, 1063–1077.

Reeves, T.G., Boundy, K.A. & Brooke, H.D. (1977), Phenological development studies with *Lupinus angustifolius* and *L. albus* in Victoria, *Australian Journal of Experimental Agriculture and Animal Husbandry* **17**, 637-644.

Siddons, P.A., Jones, R.J.A., Hollis, J.M., Hallett, S.H., Huyghe, C., Day, J.M., Scott, T. & Milford, G.F. (1994), The use of a land suitability model to predict where autumn-sown, determinate genotypes of the white lupin (*Lupinus albus*) might be grown in England and Wales, *Journal of Agricultural Science, Cambridge* **123**, 199–205.

Wiersema, J.H., Kirkbride, J.H. & Gunn, C.R. (1990), *Legume (Fabaceae) nomenclature in the USDA germplasm system*, Technical Bulletin 1757 (United States Department of Agriculture).

Williams, I.H., Martin, A.P., Ferguson, A.W. & Clark, S.J. (1990), Effect of pollination on flower, pod and seed production in white lupin (*Lupinus albus*), *Journal of Agricultural Science* **115**, 67–73.

Zadoks, J.C., Chang, T.T. & Konzak, C.F. (1974), A decimal code for the growth stages of cereals, *Weed Research* **14**, 415–421.

INDEX

Note: because Narrow-leafed Lupin is discussed throughout the text it is not indexed.

A
abscission, 16, 33, 38, 39, 41, 42, 62
adaptation, 18, 37, 39, 49
agrochemicals, xiii, 3, 55
Albus Lupin, 5, 7 12, 17, 18, 20, 22, 28, 41, 51, 52, 67–72, 77
alkaloid, 8
anchorage, 9
anthesis, 2, 3, 23, 37–42, 62, 78
    timing, 25–27, 32, 34, 37, 39, 41, 42
architecture, 26–30
    determinate, 28, 30
    indeterminate, 28, 29

B
bees, 41
bract, 32, 33–35, 38, 39, 78
Bradyrhizobium lupini, 52
branch, 19
    apex, 21, 25
    apical, 22
    basal, 22–24
    bud, 19, 22, 25–27
    first-order, 22–26
    flower initiation, 2, 26, 27
    flowering, 2, 26
    growth, 24, 25, 41
    lower, 22
    meristem, 25, 26
    mid-stem, 22, 23
    nomenclature, 23, 24, 28
    second-order, 22, 24–26
    structure, 21
    third-order, 22, 24–26
    upper, 22–27
    vigour, 22
branches
    pod set, growth, 27, 48
    seed growth, 48
branching, 21–30
    determinate, 28, 30
    epigonal, 28–30
    fully restricted, 29
    indeterminate, 28, 29, 32
    mildly restricted, 29
    morphology, 21–26
    pattern, 22–24
    reduced, 28, 30
    restricted, 28, 29, 42, 44
bud
    inflorescence, 13, 25, 26, 59, 62, 80
    lateral, 19, 21, 25
    terminal, 6, 15, 26, 59, 62

C
calyx, 34, 35, 38
canopy, 27, 44, 53
chickpea, 7
compaction, 11, 50, 51, 54
competition, 27, 33, 43, 44
corolla, 35, 38, 39
cotyledons, 6, 7, 9, 24, 47, 59
    colour, 47, 48, 63
    mass, 7
crop
    comparison, 55
    development, 57, 66, 73, 79
    emergence, 59
    flowering, 62
    management, 3, 4, 74, 76
    monitoring, 55–64, 67, 74, 76
cytokinin, 42

D
daylength, 3, 17, 18, 25, 28, 30, 32, 39, 66, 70
development, 55–72
    assessing, 55–64, 73–81
    gradient, 56
    phase, 1, 3
    predicting, 65–72
    rate, 1, 43
    scale, 55–64
    stage, 1, 3, 55–64, 73, 74, 79
dissection, 56, 67, 73, 77, 78
drought, 17–20, 37, 43
    timing, 43

E
embryo, 6, 45, 47, 63, 78
    damage, 6, 7
    development, 47, 48
    growth, 45–47
    root, 6, 47
    shoot, 6, 47
emergence (seedling), 1–12, 59, 67, 73–75, 80
    calculating, 11, 75, 80, 81
    crop, 11, 59
    definition, 10, 11, 58, 59
    epigeal, 9
    hypogeal, 9
    rate, 11, 12, 68
    recording, 73, 75
    resistance, 9
    timing, 12, 13, 59, 67, 68, 73
endosperm, 45–47, 63, 78
epicotyl, 6
epigeal emergence, 9
establishment, 7

INDEX

F
faba bean, xiv, 7, 9, 56
Fabaceae, 40, 41
Faboideae, xiv
fertilisation, 37, 40, 44
field pea, xiv, 7, 9
floral diagram, 35
floral initiation, 2, 3, 16, 24–27, 31, 32, 62, 68, 71, 73, 74, 77
floral phase, 2, 27, 31, 32, 78
flower, 31–36, 39, 68
    abscission, 38, 41, 42, 62
    development, 31–36, 38, 39, 56, 62
    examination, 78
    formation, 3
    initiation, 13, 24, 26, 31, 32, 66
    open, 39, 56, 62
    opening, 39, 62
    primordia, 32, 33
    senescence, 62
    shedding, 41, 42
flowering, 4, 28, 31, 37–42, 62, 74, 78, 80
    complete, 54, 80
    recording, 39, 73, 80
    timing, 4, 26, 32, 37, 39, 40, 70, 71, 73, 76, 79, 80
frost, 44, 48
funicle, 6, 45

G
germination, 8, 9, 11, 50, 54, 57, 79
    definition, 9, 59
    rate, 11, 66
growth scale, 55–64
growth stages, 57–64

H
hard seededness, 8
harvest maturity, 2, 37, 44, 47, 56, 64, 71, 75
hilum, 6
hypocotyl, 6, 9, 10, 47, 51, 57, 59, 76
    elongation, 9, 11, 59
    hook, 9, 10, 57
hypogeal emergence, 9

I
imbibition, 8, 78
inflorescence, 23, 25, 26, 29, 31–36, 39, 62, 63
    bud, 13, 23, 25, 26, 59, 62, 80
    bud visible, 25, 61, 62
    cyme, 28, 29
    determinate, 28, 29
    development, 63
    indeterminate, 29, 33
    internode, 33
inoculation, 52
internode, 24, 25
    growth, 24, 59
    length, 24, 25, 59

L
leaf, 10, 18, 19
    area, 18, 37, 57, 66
    arrangement, 14, 19, 20
    axil, 19, 21
    development, 13–20, 62
    drop, 54
    emergence, 16, 17, 57, 59, 66, 69, 70
    initiation, 13, 16, 17, 21, 22, 26, 27, 32, 68, 69, 72
    morphology, 19
    movement, 20
    number, 16–18, 21, 22, 26–28, 32, 66–68, 71, 72, 77, 78
    opening, 62
    primordia, 6, 13–15, 26, 32, 33, 47, 68, 72
    senescence, 16
    shading, 20, 27
    size, 19
    solar tracking, 20
    structure, 19
leaflet, 14, 15, 19, 59
    length, 16, 18, 19
    width, 19
legume, xiv, 45
lentil, xiv
life cycle, 1–4, 27, 56
light, 62, 66
    low, 37, 59
lodging, 71
Lupinus
    albus, vii, xiv, 6, 7
    angustifolius, vii, xiv, 6, 7
    cosentinii, vii, 8, 51
    luteus, xiv, 6, 7
    pilosus, 51

M
maturity, 56, 63, 64, 74, 75, 80
meristematic dome, 6, 14, 15, 26, 32, 47
microscope, 77
model, 4, 65–81
    computer, 4
    crop, 65
    physiological, 65
modelling, 65–72
mucigel, 50

N
nitrogen fixation, xiv, 53
nodulation, 52, 53, 76
nodules, 49, 51–53
nutrient deficiency, 17, 18, 54
nutrition, 22, 30, 50, 51, 53, 66

O
ovary, 33, 34, 36, 40, 44, 78
    growth, 41, 42

P
pea, xiv, 7, 9
pedicel, 35, 38, 41, 42
peduncle, 25, 33, 61
petal, 33–35, 38, 45, 78
    keel, 34–36, 40, 41
    senescence, 38, 39
    standard, 34, 35, 38, 39
    wing, 34–36, 38
petiole, 15, 19
    length, 16, 18, 19
photography, 73, 79
photosynthate, 53, 54
phyllochron, 17
physiological maturity, 2, 16, 47, 56, 63, 64, 75, 80
plant density, 22, 25, 30, 62, 66
plant height, 25, 54
plant sampling, 75, 76
plastochron, 17, 22
pod, 38, 43–48
    abortion, 41, 42, 44
    abscission, 44, 62
    colour, 45, 63, 78
    cross-walls, 45, 48, 63, 78
    development, 43–48, 56, 63
    examination, 78
    filling, 37, 39, 43, 44
    growth, 2, 28, 32, 37, 44, 45, 48

lengthening, 45
number, 40
septa, 45, 48, 62, 63, 78
set, 32, 38, 39, 41, 42, 44, 62
shattering, 45
thickening, 45
wall, 45
pollen, 40, 41
    germination, 40
    tube, 40
pollination, 37, 44
    cross, 40, 41
    self, 40, 41
primordia, 32–34
    bract, 32
    branch, 26
    flower, 31–34
    leaf, 6, 13–15, 17, 26, 32, 33, 68, 72
    stipule, 14, 15, 32
pulvinus, 16, 19, 20

Q
quadrat, 75, 76, 80, 81

R
raceme, 29, 32, 62
rachis, 25, 33–35, 39
radicle, 6, 8, 9, 47, 50, 54, 57
    elongation, 8, 11
    pocket, 6, 47
radicle elongation, 11
rainfall records, 79
rhizobia, 52
root, 9, 49–54, 76
    abundance, 53
    adventitious, 51, 60
    apex, 50
    branching, 51
    cap, 50
    cluster, 51, 52
    distribution, 53
    elongation, 50, 54
    first-order, 50–52
    growth, 53, 54
    hairs, 50
    inspection, 76
    lateral, 10, 50–52
    length, 54
    proportion of plant, 54
    proteoid, 51, 52
    tap, 10, 51–53, 57
    tip, 50
    weight, 54

S
sampling, 73–76, 80
sand blasting, 13
scale, 3
seed, 5–12, 43–48
    abortion, 42, 48
    coat, 5–7, 47, 48, 57, 63, 77, 78
    colour, 6, 47, 48, 63, 78
    composition, 7, 8
    damage, 6, 7, 9
    development, 43–48, 56, 63
    examination, 78
    filling, 37, 42, 44, 47
    funicle, 6, 45
    growth, 2, 28, 32, 45, 48, 63
    hilum, 6
    hull, 5
    imbibition, 8
    mass, 5, 7
    quality, 67, 74
    sampling, 74
    shape, 5
    size, 12, 68
    stalk, 6, 45, 48, 64
    structure, 5
    water content, 5, 47, 63
    weight, 6, 7, 47, 63, 74
seed testing, 74
seedling, 10, 14, 50, 57, 59, 60
    axis, 5, 6
    growth, 7
    vigour, 12, 74
sepal, 33–35, 45, 78
septa, 45, 63, 78
shading, 19, 22, 27, 44
shoot apex, 6, 10, 13–16, 21, 26, 32, 47, 67, 77, 78
silly seedling, 9
soil, 9, 11
    compaction, 11, 50, 51, 53, 75
    crusting, 9, 10
    moisture, 11, 53, 54
    pH, 72
    strength, 11
    temperature, 11, 12
sowing
    date, 25, 30, 32, 39, 40, 66, 70, 71, 79
    depth, 10–12, 51, 67, 76
    dry, 11, 79
    timing, 2, 18, 53
soya bean, 7, 8, 56
stamens, 33–36, 40, 78
    anthers, 34–36, 40, 78
    filaments, 36, 40, 41
    tube, 34–36, 45
stem, 21–30
    elongation, 2, 24, 59, 60, 62
    length, 24
stigma, 34–36, 40, 41, 45
stipule, 10, 16, 19
    primordia, 14, 15, 32
storage of plant samples, 76
style, 34–36, 40, 45

T
temperature, 3, 17, 18, 32, 37, 39, 53, 59, 66–70, 79
    high, 11, 37, 44, 48, 79
    low, 11, 37, 68, 79
    maximum, 67–69, 71, 79, 81
    mean, 68
    minimum, 67–71, 79, 81
    optimum, 68
    soil, 11, 12
testa, 5, 6
thermal time, 3, 12, 17, 66–70, 72, 80, 81
time of sowing, 18, 53

V
variation
    between plants, 75
vernalisation, 18, 39, 68, 69, 71, 72
vegetative phase, 2, 31, 27, 77

W
water use efficiency, 20
waterlogging, 11, 44, 50, 51, 53, 76
weather data, 73, 79

Y
Yellow Lupin, xiv, 5, 7, 12, 17, 41, 51, 54
yield, 43, 44

Z
Zadoks scale, 56